nec2 and nec3 compared

ROBERT GERRARD

NEC is a division of Thomas Telford Ltd, which is a wholly owned subsidiary of the Institution of Civil Engineers (ICE).

The NEC is a family of standard contracts, each of which has these characteristics:

- Its use stimulates good management of the relationship between the two parties to the contract and, hence, of the work included in the contract.
- It can be used in a wide variety of commercial situations, for a wide variety of types of work and in any location.
- It is a clear and simple document – using language and a structure which are straightforward and easily understood.

The NEC3 suite of contracts comprises 23 documents available in both print and electronic formats. Also available in addition to this publication is a range of support publications.

Full information on all publications and other NEC products and services is available at www.neccontract.com

A catalogue record for this book is available from the British Library

ISBN 0 7277 3384 2

© nec 2005

9 8 7 6 5 4 3 2 1

Disclaimer

Typeset by Academic + Technical, Bristol, UK

Printed and bound in Great Britain by Bell & Bain Limited, Glasgow, UK

NEC2 and NEC3 Compared is a clause-by-clause comparison of the changes between the second (NEC2) and third (NEC3) editions of the *NEC Engineering and Construction Contract* (ECC). It presents the wording of ECC2 against that of ECC3. All information that has been deleted from ECC2 can be found on the left hand side shaded in grey. Additional information or substantial changes that are new to ECC3 are on the right highlighted in red. For completeness, the final version of ECC2 together with Y(UK)2 and Y(UK)3 have been used for a direct comparison. Please note that ECC2 has been re-ordered to correspond directly with the new structure of ECC3. There are additional explanatory notes under appropriate clauses in ECC3, giving details of more significant changes. These can be found in red shaded boxes. The index is that of ECC3.

Those already using NEC2 will want to know what the scale of changes are and how these will affect them and the implications of using NEC3. The principle behind the book is that the reader will be able to identify at first glance all information that is new and what has been deleted between the two editions.

NEC3 comprises a suite of entirely consistent documents. The guide covers in detail the differences between the *NEC2 Engineering and Construction Contract* and the *NEC3 Engineering and Construction Contract.* However, principles of the changes it introduces have been applied across the complete suite of NEC3 contracts. So anybody wishing to understand the new provisions applied across the suite of NEC3 contracts, will benefit from this book.

The NEC has been extensively used in construction and engineering in both the United Kingdom and elsewhere since its launch in 1995. Since 1995, the NEC Panel (the body within the Institution of Civil Engineers responsible for the NEC) has received feedback on the use of NEC2. In 2002 the NEC Panel decided to undertake a formal review of the NEC incorporating feedback received to date and inviting feedback from the NEC Users' Group and the construction industry at large.

There was considerable feedback and to ensure a comprehensive review of all users' contributions, a number of working groups were set up to appraise comments and recommend actions. The NEC Panel then debated the principles of change and concluded that the time was right to produce updated versions of existing NEC2 documents together with new forms of contract to cover term service contracts and framework contracts. These have now been published as the NEC3 suite of documents.

At the NEC Users' Group annual seminar in 2004 the UK Office of Government Commerce (OGC) delivered their keynote address and commented that NEC was reasonably *Achieving Excellence in Construction* (AEC) compliant. AEC is a UK government initiative and through this, public sector clients commit to maximise, by continuous improvement, the efficiency, effectiveness and value for money of their procurement of new works, maintenance and refurbishment. Debate followed with the OGC in how NEC could be updated to become fully compliant with AEC principles without compromising the values that NEC stands for, together with the premise that NEC is for use with both public and private sector work, in any sector within the engineering and construction industries, and in any part of the World.

As a result of this dialog the NEC3 suite of documents is now endorsed by the OGC as follows: 'OGC advises public sector procurers that the form of contract used has to be selected according to the objectives of the project, aiming to satisfy the *Achieving Excellence in Construction (AEC)* principles.

This edition of the NEC (NEC3) complies fully with the AEC principles. OGC recommends the use of NEC3 by public sector construction procurers on their construction projects.'

CONTENTS

CONTENTS

CONTENTS

SCHEDULE OF OPTIONS

The strategy for choosing the form of contract starts with a decision between six main options, one of which must be chosen.

Option A	Priced contract with activity schedule
Option B	Priced contract with bill of quantities
Option C	Target contract with activity schedule
Option D	Target contract with bill of quantities
Option E	Cost reimbursable contract
Option F	Management contract

The following secondary options should then be considered. It is not necessary to use any of them. Any combination other than those stated may be used.

Option N	Price adjustment for inflation (not to be used with Options E and F)
Option T	Changes in the law
Option K	Multiple currencies (not to be used with Options C, D, E and F)
Option H	Parent company guarantee
Option L	Sectional Completion
Option Q	Bonus for early Completion
Option R	Delay damages
Option X12	The NEC Partnering Option
Option G	Performance bond
Option J	Advanced payment to the *Contractor*
Option M	Limitation of the *Contractor*'s liability for his design to reasonable skill and care
Option P	Retention (not to be used with Option F)
Option S	Low performance damages
Option U	The Construction (Design and Management) Regulations 1994 (to be used for contracts in the UK)
Option V	Trust Fund
Y(UK)2	The Housing Grants, Construction and Regeneration Act 1996
Y(UK)3	The Contracts (Rights of Third Parties) Act 1999
Option Z	Additional conditions of contract

SCHEDULE OF OPTIONS

The strategy for choosing the form of contract starts with a decision between six main Options, one of which must be chosen.

Option A	Priced contract with activity schedule
Option B	Priced contract with bill of quantities
Option C	Target contract with activity schedule
Option D	Target contract with bill of quantities
Option E	Cost reimbursable contract
Option F	Management contract

One of the following dispute resolution Options must be selected to complete the chosen main Option.

Option W1	Dispute resolution procedure (used unless the United Kingdom Housing Grants, Construction and Regeneration Act 1996 applies).
Option W2	Dispute resolution procedure (used in the United Kingdom when the Housing Grants, Construction and Regeneration Act 1996 applies).

The following secondary Options should then be considered. It is not necessary to use any of them. Any combination other than those stated may be used.

Option X1	Price adjustment for inflation (used only with Options A, B, C and D)
Option X2	Changes in the law
Option X3	Multiple currencies (used only with Options A and B)
Option X4	Parent company guarantee
Option X5	Sectional Completion
Option X6	Bonus for early Completion
Option X7	Delay damages
Option X12	Partnering
Option X13	Performance bond
Option X14	Advanced payment to the *Contractor*
Option X15	Limitation of the *Contractor*'s liability for his design to reasonable skill and care
Option X16	Retention (not used with Option F)
Option X17	Low performance damages
Option X18	Limitation of liability
Option X20	Key Performance Indicators (not used with Option X12)

The following Options dealing with national legislation should be included if required.

Option Y(UK)2	The Housing Grants, Construction and Regeneration Act 1996
Option Y(UK)3	The Contracts (Rights of Third Parties) Act 1999
Option Z	*Additional conditions of contract*
Note	Options X8 to X11, X19 and Y(UK)1 are not used.

THE NEC ENGINEERING AND CONSTRUCTION CONTRACT

CORE CLAUSES

1 General

Actions **10**

10.1   The *Employer*, the *Contractor*, the *Project Manager* and the *Supervisor* shall act as stated in this contract and in a spirit of mutual trust and co-operation. The *Adjudicator* shall act as stated in this contract and in a spirit of independence.

Identified and defined terms **11**

11.1   In these conditions of contract, terms identified in the Contract Data are in italics and defined terms have capital initials.

11.2   (14) The Accepted Programme is the programme identified in the Contract Data or is the latest programme accepted by the *Project Manager*. The latest programme accepted by the *Project Manager* supersedes previous Accepted Programmes.

   (13) Completion is when the *Contractor* has

   - done all the work which the Works Information states he is to do by the Completion Date and

   - corrected notified Defects which would have prevented the *Employer* from using the *works*.

   (12) The Completion Date is the *completion date* unless later changed in accordance with this contract.

   (3) The Contract Date is the date when this contract came into existence.

   (15) A Defect is

   - a part of the *works* which is not in accordance with the Works Information or

   - a part of the *works* designed by the *Contractor* which is not in accordance with

     - the applicable law or

     - the *Contractor*'s design which has been accepted by the *Project Manager*.

⏺nec Engineering and Construction Contract

CORE CLAUSES

1 General

Actions 10

10.1   The *Employer*, the *Contractor*, the *Project Manager* and the *Supervisor* shall act as stated in this contract and in a spirit of mutual trust and co-operation.

Identified and defined 11
terms 11.1   In these conditions of contract, terms identified in the Contract Data are in italics and defined terms have capital initials.

There are two new defined terms, and some of the others have been amended. The subclauses in 11.2 have been re-arranged alphabetically.

11.2   (1) The Accepted Programme is the programme identified in the Contract Data or is the latest programme accepted by the *Project Manager*. The latest programme accepted by the *Project Manager* supersedes previous Accepted Programmes.

Where the Works Information is silent on what work the *Contractor* is to do by the Completion Date, a new test has been added. In this case the test is whether the *Contractor* has done all the work necessary for the *Employer* to use the *works* and for Others to do their work.

(2) Completion is when the *Contractor* has

- done all the work which the Works Information states he is to do by the Completion Date and
- corrected notified Defects which would have prevented the *Employer* from using the *works* and Others from doing their work.

If the work which the *Contractor* is to do by the Completion Date is not stated in the Works Information, Completion is when the *Contractor* has done all the work necessary for the *Employer* to use the *works* and for Others to do their work.

(3) The Completion Date is the *completion date* unless later changed in accordance with this contract.

(4) The Contract Date is the date when this contract came into existence.

(5) A Defect is

- a part of the *works* which is not in accordance with the Works Information or
- a part of the *works* designed by the *Contractor* which is not in accordance with the applicable law or the *Contractor*'s design which the *Project Manager* has accepted.

(16) The Defects Certificate is either a list of Defects that the *Supervisor* has notified before the *defects date* which the *Contractor* has not corrected or, if there are no such Defects, a statement that there are none.

(11) Equipment is items provided by the *Contractor* and used by him to Provide the Works and which the Works Information does not require him to include in the *works*.

(17) The Fee is the amount calculated by applying the *fee percentage* to the amount of Actual Cost.

(2) Others are people or organisations who are not the *Employer*, the *Project Manager*, the *Supervisor*, the *Adjudicator*, the *Contractor*, or any employee, Subcontractor or supplier of the *Contractor*.

(1) The Parties are the *Employer* and the *Contractor*.

(10) Plant and Materials are items intended to be included in the *works*.

(4) To Provide the Works means to do the work necessary to complete the *works* in accordance with this contract and all incidental work, services and actions which this contract requires.

(7) The Site is the area within the *boundaries of the site* and the volumes above and below it which are affected by work included in this contract.

(6) The Defects Certificate is either a list of Defects that the *Supervisor* has notified before the *defects date* which the *Contractor* has not corrected or, if there are no such Defects, a statement that there are none.

(7) Equipment is items provided by the *Contractor* and used by him to Provide the Works and which the Works Information does not require him to include in the *works*.

(8) The Fee is the sum of the amounts calculated by applying the *subcontracted fee percentage* to the Defined Cost of subcontracted work and the *direct fee percentage* to the Defined Cost of other work.

In ECC2 there was provision for a Fee that, when applicable, was calculated by applying a single *fee percentage* to Actual Cost. This is now replaced by two fee percentages, one applied to the Defined Cost of subcontracted work and another to the Defined Cost of other work. The Fee has different purposes in the main Options. This change, together with amendments to the definition of a Subcontractor, Defined Cost and the Schedules of Cost Components, should remove an area of uncertainty within ECC2 when dealing with Subcontractor's costs in particular.

(9) A Key Date is the date by which work is to meet the Condition stated. The Key Date is the *key date* stated in the Contract Data and the Condition is the *condition* stated in the Contract Data unless later changed in accordance with this contract.

The introduction of 'Key Dates' is a new feature of ECC3. This is not a part of the *works* that contractually is described as a *section*, but instead could be a key interface point in a project. This may be between two trades or when the *Contractor* must achieve a certain deliverable to suit the *Employer*'s procurement of related aspects. The use of Key Dates is entirely optional. Where this is incorporated, both the Key Date and the Condition must be clearly stated in the Contract Data.

(10) Others are people or organisations who are not the *Employer,* the *Project Manager*, the *Supervisor*, the *Adjudicator*, the *Contractor* or any employee, Subcontractor or supplier of the *Contractor*.

(11) The Parties are the *Employer* and the *Contractor*.

(12) Plant and Materials are items intended to be included in the *works*.

(13) To Provide the Works means to do the work necessary to complete the *works* in accordance with this contract and all incidental work, services and actions which this contract requires.

(14) The Risk Register is a register of the risks which are listed in the Contract Data and the risks which the *Project Manager* or the *Contractor* has notified as an early warning matter. It includes a description of the risk and a description of the actions which are to be taken to avoid or reduce the risk.

To reinforce the pro-active approach that NEC contracts take to dealing with risk, the ECC now includes provisions for managing project risk through a Risk Register. This document is incorporated into the contract through the Contract Data. Occurrences of such risks included or others that arise during the currency of the contract do not constitute a compensation event. The emphasis is instead clearly on notifying risks as early warnings, adding these to the Risk Register and using the risk reduction meeting to avoid or reduce such risks. The Risk Register therefore introduces a contractual process for managing risk; the contractual responsibility for risk itself remains as provided for in clauses 60 and 80.

(15) The Site is the area within the *boundaries of the site* and the volumes above and below it which are affected by work included in this contract.

(6) Site Information is information which

- describes the Site and its surroundings and
- is in the documents which the Contract Data states it is in.

(9) A Subcontractor is a person or corporate body who has a contract with the *Contractor* to provide part of the *works* or to supply Plant and Materials which he has wholly or partly designed specifically for the *works*.

(8) The Working Areas are the *working areas* unless later changed in accordance with this contract.

(5) Works Information is information which either

- specifies and describes the *works* or
- states any constraints on how the *Contractor* Provides the Works

and is either

- in the documents which the Contract Data states it is in or
- in an instruction given in accordance with this contract.

Interpretation and the Law   **12**

12.1   In this contract, except where the context shows otherwise, words in the singular also mean in the plural and the other way round and words in the masculine also mean in the feminine and neuter.

12.2   This contract is governed by the *law of the contract*.

(16) Site Information is information which

- describes the Site and its surroundings and
- is in the documents which the Contract Data states it is in.

(17) A Subcontractor is a person or organisation who has a contract with the *Contractor* to

- construct or install part of the *works*,
- provide a service necessary to Provide the Works or
- supply Plant and Materials which the person or organisation has wholly or partly designed specifically for the *works*.

The definition of a Subcontractor has been extended to include those persons or organisations providing a service. This service must, however, be necessary to Provide the Works and could include the likes of design. It is likely that additional persons or organisations will fall within the definition of Subcontractor than those under ECC2 which has implications not least in the payment provisions and assessment of compensation events.

(18) The Working Areas are those parts of the *working areas* which are

- necessary for Providing the Works and
- used only for work in this contract

unless later changed in accordance with this contract.

The definition of Working Areas has been narrowed to exclude *working areas* that are not necessary to Provide the Works or are used for works other than in this contract. This tightens the drafting and will therefore generally exclude the *Contractor*'s head office from falling within this definition.

(19) Works Information is information which either

- specifies and describes the *works* or
- states any constraints on how the *Contractor* Provides the Works

and is either

- in the documents which the Contract Data states it is in or
- in an instruction given in accordance with this contract.

Interpretation and the law   **12**

12.1   In this contract, except where the context shows otherwise, words in the singular also mean in the plural and the other way round and words in the masculine also mean in the feminine and neuter.

12.2   This contract is governed by the *law of the contract*.

12.3   No change to this contract, unless provided for by the *conditions of contract*, has effect unless it has been agreed, confirmed in writing and signed by the Parties.

A new subclause confirms a three-stage test for any potential changes to the existing terms of a contract. Changes must be agreed, confirmed in writing and signed by the Parties. Therefore as always intended, the *Project Manager* cannot change the terms of a contract.

12.4   This contract is the entire agreement between the Parties.

A new subclause to clarify that the Parties cannot later rely upon documents not incorporated in the contract.

Communications **13**

13.1 Each instruction, certificate, submission, proposal, record, acceptance, notifica tion and reply which this contract requires is communicated in a form which ca be read, copied and recorded. Writing is in the *language of this contract*.

13.2 A communication has effect when it is received at the last address notified b the recipient for receiving communications or, if none is notified, at the addres of the recipient stated in the Contract Data.

13.3 If this contract requires the *Project Manager*, the *Supervisor* or the *Contractor* t reply to a communication, unless otherwise stated in this contract, he replie within the *period for reply*.

13.4 The *Project Manager* replies to a communication submitted or resubmitted t him by the *Contractor* for acceptance. If his reply is not acceptance, he states hi reasons and the *Contractor* resubmits the communication within the *period fo reply* taking account of these reasons. A reason for withholding acceptance i that more information is needed in order to assess the *Contractor*'s submissio fully.

13.5 The *Project Manager* may extend the *period for reply* to a communication if th *Project Manager* and the *Contractor* agree to the extension before the reply i due. The *Project Manager* notifies the extension which has been agreed to th *Contractor*.

13.6 The *Project Manager* issues his certificates to the *Employer* and the *Contracto* The *Supervisor* issues his certificates to the *Project Manager* and the *Contractor*.

13.7 A notification which this contract requires is communicated separately fror other communications.

13.8 The *Project Manager* may withhold acceptance of a submission by the *Contrac tor*. Withholding acceptance for a reason stated in this contract is not a compen sation event.

The *Project Manager* and **14**
the *Supervisor* 14.1 The *Project Manager*'s or the *Supervisor*'s acceptance of a communication fror the *Contractor* or of his work does not change the *Contractor*'s responsibility t Provide the Works or his liability for his design.

14.2 The *Project Manager* and the *Supervisor*, after notifying the *Contractor*, ma delegate any of their actions and may cancel any delegation. A reference to a action of the *Project Manager* or the *Supervisor* in this contract includes a action by his delegate.

14.3 The *Project Manager* may give an instruction to the *Contractor* which change the Works Information.

14.4 The *Employer* may replace the *Project Manager* or the *Supervisor* after he ha notified the *Contractor* of the name of the replacement.

Communications **13**

    13.1   Each instruction, certificate, submission, proposal, record, acceptance, notification, reply and other communication which this contract requires is communicated in a form which can be read, copied and recorded. Writing is in the *language of this contract*.

> The additional wording clarifies the intention that all communications this contract requires must be in a form that can be read, copied and recorded.

    13.2   A communication has effect when it is received at the last address notified by the recipient for receiving communications or, if none is notified, at the address of the recipient stated in the Contract Data.

    13.3   If this contract requires the *Project Manager*, the *Supervisor* or the *Contractor* to reply to a communication, unless otherwise stated in this contract, he replies within the *period for reply*.

    13.4   The *Project Manager* replies to a communication submitted or resubmitted to him by the *Contractor* for acceptance. If his reply is not acceptance, the *Project Manager* states his reasons and the *Contractor* resubmits the communication within the *period for reply* taking account of these reasons. A reason for withholding acceptance is that more information is needed in order to assess the *Contractor*'s submission fully.

    13.5   The *Project Manager* may extend the *period for reply* to a communication if the *Project Manager* and the *Contractor* agree to the extension before the reply is due. The *Project Manager* notifies the *Contractor* of the extension which has been agreed.

    13.6   The *Project Manager* issues his certificates to the *Employer* and the *Contractor*. The *Supervisor* issues his certificates to the *Project Manager* and the *Contractor*.

    13.7   A notification which this contract requires is communicated separately from other communications.

    13.8   The *Project Manager* may withhold acceptance of a submission by the *Contractor*. Withholding acceptance for a reason stated in this contract is not a compensation event.

The *Project Manager* and **14**
the *Supervisor* 14.1   The *Project Manager*'s or the *Supervisor*'s acceptance of a communication from the *Contractor* or of his work does not change the *Contractor*'s responsibility to Provide the Works or his liability for his design.

    14.2   The *Project Manager* and the *Supervisor*, after notifying the *Contractor*, may delegate any of their actions and may cancel any delegation. A reference to an action of the *Project Manager* or the *Supervisor* in this contract includes an action by his delegate.

    14.3   The *Project Manager* may give an instruction to the *Contractor* which changes the Works Information or a Key Date.

> The *Project Manager* is now empowered to give instructions to the *Contractor* to change any Key Dates as well as the Works Information.

    14.4   The *Employer* may replace the *Project Manager* or the *Supervisor* after he has notified the *Contractor* of the name of the replacement.

Adding to the *working areas* **15**

15.1   The *Contractor* may submit a proposal for adding to the Working Areas to the *Project Manager* for acceptance. A reason for not accepting is that

- the proposed addition is not necessary for Providing the Works or
- the proposed area will be used for work not in this contract.

Early warning **16**

16.1   The *Contractor* and the *Project Manager* give an early warning by notifying the other as soon as either becomes aware of any matter which could

- increase the total of the Prices,
- delay Completion or
- impair the performance of the *works* in use.

16.2   Either the *Project Manager* or the *Contractor* may instruct the other to attend an early warning meeting. Each may instruct other people to attend if the other agrees.

16.3   At an early warning meeting those who attend co-operate in

- making and considering proposals for how the effect of each matter which has been notified as an early warning can be avoided or reduced,
- seeking solutions that will bring advantage to all those who will be affected and
- deciding upon actions which they will take and who, in accordance with this contract, will take them.

---

www.neccontract.com

Adding to the Working Areas   **15**

15.1   The *Contractor* may submit a proposal for adding an area to the Working Areas to the *Project Manager* for acceptance. A reason for not accepting is that the proposed area is either not necessary for Providing the Works or used for work not in this contract.

> The amendments to this subclause line up with the revised subclause 11.2(18). The *Contractor* is able to submit a proposal for adding an area to the Working Areas but it must satisfy a two-stage test and the *Project Manager* has the right not to accept this proposal should one of the tests fail.

Early warning   **16**

16.1   The *Contractor* and the *Project Manager* give an early warning by notifying the other as soon as either becomes aware of any matter which could

- increase the total of the Prices,
- delay Completion,
- delay meeting a Key Date or
- impair the performance of the *works* in use.

The *Contractor* may give an early warning by notifying the *Project Manager* of any other matter which could increase his total cost. The *Project Manager* enters early warning matters in the Risk Register. Early warning of a matter for which a compensation event has previously been notified is not required.

> A new event that both the *Contractor* and the *Project Manager* must give an early warning for is any matter that could delay meeting a Key Date. Another new provision is for the *Contractor* to give an early warning at his discretion on any other matter that could increase the *Contractor*'s cost. This will certainly be of interest to the *Employer* in Options C, D and E, where this cost is likely to be borne at least in part by the *Employer*, and of course is of interest to both Parties in all Options, keeping to the open nature and partnering spirit of the contract.

16.2   Either the *Project Manager* or the *Contractor* may instruct the other to attend a risk reduction meeting. Each may instruct other people to attend if the other agrees.

> The ECC2 early warning meeting is now very usefully re-named as the risk reduction meeting to clarify the purpose of the attendees at the meeting.

16.3   At a risk reduction meeting, those who attend co-operate in

- making and considering proposals for how the effect of the registered risks can be avoided or reduced,
- seeking solutions that will bring advantage to all those who will be affected,
- deciding on the actions which will be taken and who, in accordance with this contract, will take them and
- deciding which risks have now been avoided or have passed and can be removed from the Risk Register.

> The Risk Register becomes the central risk management tool and is kept up to date by the *Project Manager*. Through early warnings, the Risk Register and the risk reduction meeting, the objective of the *Project Manager* and *Contractor* is to set about reducing or avoiding the registered risks. The Risk Register not only records current and live risks, but also records those which have been avoided or have passed. This subclause does refer to removing risks from the Risk Register, however changing the status of such risks to 'dead', for example, may be a practical alternative to ensure the record is properly kept to better inform future decision making.

16.4   The *Project Manager* records the proposals considered and the decisions take at an early warning meeting and gives a copy of his record to the *Contractor*.

Ambiguities and inconsistencies   17

17.1   The *Project Manager* or the *Contractor* notifies the other as soon as eith becomes aware of an ambiguity or inconsistency in or between the documen which are part of this contract. The *Project Manager* gives an instructio resolving the ambiguity or inconsistency.

Illegal and impossible requirements   19

19.1   The *Contractor* notifies the *Project Manager* as soon as he becomes aware th the Works Information requires him to do anything which is illegal or impossibl If the *Project Manager* agrees, he gives an instruction to change the Wor Information appropriately.

16.4   The *Project Manager* revises the Risk Register to record the decisions made at each risk reduction meeting and issues the revised Risk Register to the *Contractor*. If a decision needs a change to the Works Information, the *Project Manager* instructs the change at the same time as he issues the revised Risk Register.

The Risk Register is updated and issued by the *Project Manager* following the decisions made at the risk reduction meeting.

Ambiguities and inconsistencies

17

17.1   The *Project Manager* or the *Contractor* notifies the other as soon as either becomes aware of an ambiguity or inconsistency in or between the documents which are part of this contract. The *Project Manager* gives an instruction resolving the ambiguity or inconsistency.

Illegal and impossible requirements

18

18.1   The *Contractor* notifies the *Project Manager* as soon as he considers that the Works Information requires him to do anything which is illegal or impossible. If the *Project Manager* agrees, he gives an instruction to change the Works Information appropriately.

Prevention

19

19.1   If an event occurs which

- stops the *Contractor* completing the *works* or
- stops the *Contractor* completing the *works* by the date shown on the Accepted Programme,

and which

- neither Party could prevent and
- an experienced contractor would have judged at the Contract Date to have such a small chance of occurring that it would have been unreasonable for him to have allowed for it,

the *Project Manager* gives an instruction to the *Contractor* stating how he is to deal with the event.

This is a new clause which deals with rare events beyond the control of either Party. Rather than refer to 'force majeure', a term much used in other forms of contract, this provision tries to describe what is meant by prevention rather than rely upon a term that has many definitions. Essentially, an event that falls within this definition is something that may affect the Prices, Key Dates, Completion Date or even lead to termination. It is therefore the *Employer* that takes the risks of matters occurring within this definition, the chance of occurrence though being small. This needs to be stringent in its application with all tests being satisfied. The event must 'stop' the Contractor as described not merely delay or disrupt, neither Party could prevent it and also it must satisfy a reasonableness test.

2 The *Contractor*'s main responsibilities

Providing the Works **20**

20.1   The *Contractor* Provides the Works in accordance with the Works Informatio

The *Contractor*'s design **21**

21.1   The *Contractor* designs the parts of the *works* which the Works Informat states he is to design.

21.2   The *Contractor* submits the particulars of his design as the Works Informat requires to the *Project Manager* for acceptance. A reason for not accepting *Contractor*'s design is that

- it does not comply with the Works Information or
- it does not comply with the applicable law

The *Contractor* does not proceed with the relevant work until the *Proj Manager* has accepted his design.

21.3   The *Contractor* may submit his design for acceptance in parts if the design each part can be assessed fully.

21.4   The *Contractor* indemnifies the *Employer* against claims, compensation a costs due to the *Contractor* infringing a patent or copyright.

Using the *Contractor*'s design **22**

22.1   The *Employer* may use and copy the *Contractor*'s design for any purpose co nected with construction, use, alteration or demolition of the *works* unless oth wise stated in the Works Information and for other purposes as stated in Works Information.

Design of Equipment **23**

23.1   The *Contractor* submits particulars of the design of an item of Equipment to *Project Manager* for acceptance if the *Project Manager* instructs him to. reason for not accepting is that the design of the item will not allow the *Contr tor* to Provide the Works in accordance with

- the Works Information,
- the *Contractor*'s design which the *Project Manager* has accepted or
- the applicable law.

People **24**

24.1   The *Contractor* either employs each key person named to do the job for h stated in the Contract Data or employs a replacement person who has be accepted by the *Project Manager*. The *Contractor* submits the name, releva qualifications and experience of a proposed replacement person to the *Proj Manager* for acceptance. A reason for not accepting the person is that l relevant qualifications and experience are not as good as those of the pers who is to be replaced.

24.2   The *Project Manager* may, having stated his reasons, instruct the *Contractor* remove an employee. The *Contractor* then arranges that, after one day, t employee has no further connection with the work included in this contract.

Co-operation **25**

25.1   The *Contractor* co-operates with Others in obtaining and providing informati which they need in connection with the *works*. He shares the Working Are with Others as stated in the Works Information.

The *Contractor*'s main responsibilities

Providing the Works	**20**	
	20.1	The *Contractor* Provides the Works in accordance with the Works Information.
The *Contractor*'s design	**21**	
	21.1	The *Contractor* designs the parts of the *works* which the Works Information states he is to design.
	21.2	The *Contractor* submits the particulars of his design as the Works Information requires to the *Project Manager* for acceptance. A reason for not accepting the *Contractor*'s design is that it does not comply with either the Works Information or the applicable law.
		The *Contractor* does not proceed with the relevant work until the *Project Manager* has accepted his design.
	21.3	The *Contractor* may submit his design for acceptance in parts if the design of each part can be assessed fully.
Using the *Contractor*'s design	**22**	
	22.1	The *Employer* may use and copy the *Contractor*'s design for any purpose connected with construction, use, alteration or demolition of the *works* unless otherwise stated in the Works Information and for other purposes as stated in the Works Information.
Design of Equipment	**23**	
	23.1	The *Contractor* submits particulars of the design of an item of Equipment to the *Project Manager* for acceptance if the *Project Manager* instructs him to. A reason for not accepting is that the design of the item will not allow the *Contractor* to Provide the Works in accordance with

- the Works Information,
- the *Contractor*'s design which the *Project Manager* has accepted or
- the applicable law.

People	**24**	
	24.1	The *Contractor* either employs each key person named to do the job stated in the Contract Data or employs a replacement person who has been accepted by the *Project Manager*. The *Contractor* submits the name, relevant qualifications and experience of a proposed replacement person to the *Project Manager* for acceptance. A reason for not accepting the person is that his relevant qualifications and experience are not as good as those of the person who is to be replaced.
	24.2	The *Project Manager* may, having stated his reasons, instruct the *Contractor* to remove an employee. The *Contractor* then arranges that, after one day, the employee has no further connection with the work included in this contract.
Working with the *Employer* and Others	**25**	
	25.1	The *Contractor* co-operates with Others in obtaining and providing information which they need in connection with the *works*. He co-operates with Others and shares the Working Areas with them as stated in the Works Information.

33.2   While the *Contractor* has possession of a part of the Site, the *Employer* gives Contractor access to and use of it and the *Employer* and the *Contractor* prov facilities and services as stated in the Works Information. Any cost incurred the *Employer* as a result of the *Contractor* not providing the facilities a services he is to provide is assessed by the *Project Manager* and paid by Contractor.

Subcontracting 26

26.1   If the *Contractor* subcontracts work, he is responsible for performing this co tract as if he had not subcontracted. This contract applies as if a Subcontractc employees and equipment were the *Contractor*'s.

26.2   The *Contractor* submits the name of each proposed Subcontractor to the *Proj Manager* for acceptance. A reason for not accepting the Subcontractor is t his appointment will not allow the *Contractor* to Provide the Works. The *C* tractor does not appoint a proposed Subcontractor until the *Project Mana* has accepted him.

26.3   The *Contractor* submits the proposed conditions of contract for each subcc tract to the *Project Manager* for acceptance unless

- the NEC Engineering and Construction Subcontract or the NEC Prof sional Services Contract is to be used or
- the *Project Manager* has agreed that no submission is required.

The *Contractor* does not appoint a Subcontractor on the proposed subcontr: conditions submitted until the *Project Manager* has accepted them. A reason not accepting them is that

- they will not allow the *Contractor* to Provide the Works or
- they do not include a statement that the parties to the subcontract shall : in a spirit of mutual trust and co-operation.

25.2   The *Employer* and the *Contractor* provide services and other things as stated in the Works Information. Any cost incurred by the *Employer* as a result of the *Contractor* not providing the services and other things which he is to provide is assessed by the *Project Manager* and paid by the *Contractor.*

> This basis of this subclause was ECC2 subclause 33.2. The Works Information needs to be properly completed to give this effect, and the costs that the *Employer* suffers due to *Contractor* failure is paid by the *Contractor*. Where the *Employer* fails to do whatever is stated in the Works Information, a compensation event arises.

25.3   If the *Project Manager* decides that the work does not meet the Condition stated for a Key Date by the date stated and, as a result, the *Employer* incurs additional cost either

- in carrying out work or
- by paying an additional amount to Others in carrying out work

on the same project, the additional cost which the *Employer* has paid or will incur is paid by the *Contractor*. The *Project Manager* assesses the additional cost within four weeks of the date when the Condition for the Key Date is met. The *Employer*'s right to recover the additional cost is his only right in these circumstances.

> This new subclause brings in the effects of the failure of the *Contractor* to achieve the Condition stated for a Key Date. Both the Key Date and Condition are provided in the Contract Data part one at time of tender. The additional cost due to the failure must be incurred by the *Employer* on the same project and is not quantified at time of tender. The *Contractor* will therefore need to consider such implications both at the time of tender and during the progress of the *works* to ensure compliance. The whole idea of the new provision of Key Dates is to highlight, and work towards, achieving certain stated aspects of *works* that are key but not provided as contractual sections.

Subcontracting **26**

26.1   If the *Contractor* subcontracts work, he is responsible for Providing the Works as if he had not subcontracted. This contract applies as if a Subcontractor's employees and equipment were the *Contractor*'s.

26.2   The *Contractor* submits the name of each proposed Subcontractor to the *Project Manager* for acceptance. A reason for not accepting the Subcontractor is that his appointment will not allow the *Contractor* to Provide the Works. The *Contractor* does not appoint a proposed Subcontractor until the *Project Manager* has accepted him.

26.3   The *Contractor* submits the proposed conditions of contract for each subcontract to the *Project Manager* for acceptance unless

- an NEC contract is proposed or
- the *Project Manager* has agreed that no submission is required.

The *Contractor* does not appoint a Subcontractor on the proposed subcontract conditions submitted until the *Project Manager* has accepted them. A reason for not accepting them is that

- they will not allow the *Contractor* to Provide the Works or
- they do not include a statement that the parties to the subcontract shall act in a spirit of mutual trust and co-operation.

> An exception for submitting proposed subcontract conditions of contract has been extended to where any NEC contract is proposed, rather than the two named NEC contracts in ECC2. This reflects the multitude of NEC contracts available to be used as a subcontract document.

Approval from Others	**27**
	27.1 The *Contractor* obtains approval of his design from Others where necessary.

Access to the work	**28**
	28.1 The *Contractor* provides access to work being done and to Plant and Mater being stored for this contract for

- the *Project Manager*,
- the *Supervisor* and
- others notified to him by the *Project Manager*.

Instructions	**29**
	29.1 The *Contractor* obeys an instruction which is in accordance with this cont and is given to him by the *Project Manager* or the *Supervisor*.

Health and safety	**18**
	18.1 The *Contractor* acts in accordance with the health and safety requireme stated in the Works Information.

Other responsibilities **27**

27.1   The *Contractor* obtains approval of his design from Others where necessary.

27.2   The *Contractor* provides access to work being done and to Plant and Materials being stored for this contract for

- the *Project Manager*,
- the *Supervisor* and
- Others notified to him by the *Project Manager*.

27.3   The *Contractor* obeys an instruction which is in accordance with this contract and is given to him by the *Project Manager* or the *Supervisor*.

27.4   The *Contractor* acts in accordance with the health and safety requirements stated in the Works Information.

These four subclauses have been collected under a new side heading 'Other responsibilities' which better describes these provisions. These were subclauses 27.1, 28.1, 29.1 and 18.1 in ECC2.

3 Time

Starting and Completion **30**

30.1   The *Contractor* does not start work on the Site until the first *possession date* and does the work so that Completion is on or before the Completion Date.

30.2   The *Project Manager* decides the date of Completion. The *Project Manager* certifies Completion within one week of Completion.

The programme **31**

31.1   If a programme is not identified in the Contract Data, the *Contractor* submits a first programme to the *Project Manager* for acceptance within the period stated in the Contract Data.

31.2   The *Contractor* shows on each programme which he submits for acceptance

- the *starting date*, *possession dates* and Completion Date,
- planned Completion,
- the order and timing of
    - the operations which the *Contractor* plans to do in order to Provide the Works and
    - the work of the *Employer* and Others either as stated in the Works Information or as later agreed with them by the *Contractor*,
- the dates when the *Contractor* plans to complete work needed to allow the *Employer* and Others to do their work,
- provisions for
    - float,
    - time risk allowances,
    - health and safety requirements and
    - the procedures set out in this contract.
- the dates when, in order to Provide the Works in accordance with his programme, the *Contractor* will need
    - possession of a part of the Site if later than its *possession date*,
    - acceptances and
    - Plant and Materials and other things to be provided by the *Employer* and
- for each operation, a method statement which identifies the Equipment and other resources which the *Contractor* plans to use,
- other information which the Works Information requires the *Contractor* show on a programme submitted for acceptance.

Time

Starting, Completion and Key Dates **30**

30.1 The *Contractor* does not start work on the Site until the first *access date* and does the work so that Completion is on or before the Completion Date.

30.2 The *Project Manager* decides the date of Completion. The *Project Manager* certifies Completion within one week of Completion.

30.3 The *Contractor* does the work so that the Condition stated for each Key Date is met by the Key Date.

> The basic obligation for the *Contractor* to meet the Condition stated for each Key Date is the basis of this new subclause.

The programme **31**

31.1 If a programme is not identified in the Contract Data, the *Contractor* submits a first programme to the *Project Manager* for acceptance within the period stated in the Contract Data.

31.2 The *Contractor* shows on each programme which he submits for acceptance

- the *starting date*, *access dates*, Key Dates and Completion Date,
- planned Completion,
- the order and timing of the operations which the *Contractor* plans to do in order to Provide the Works,
- the order and timing of the work of the *Employer* and Others as last agreed with them by the *Contractor* or, if not so agreed, as stated in the Works Information,
- the dates when the *Contractor* plans to meet each Condition stated for the Key Dates and to complete other work needed to allow the *Employer* and Others to do their work,
- provisions for

  - float,
  - time risk allowances,
  - health and safety requirements and
  - the procedures set out in this contract,

- the dates when, in order to Provide the Works in accordance with his programme, the *Contractor* will need

  - access to a part of the Site if later than its *access date*,
  - acceptances,
  - Plant and Materials and other things to be provided by the *Employer* and
  - information from Others,

- for each operation, a statement of how the *Contractor* plans to do the work identifying the principal Equipment and other resources which he plans to use and
- other information which the Works Information requires the *Contractor* to show on a programme submitted for acceptance.

> In each programme, the *Contractor* now shows any stated Key Dates, and also *access dates* instead of *possession dates* in ECC2. For each operation, the programme needs to include a statement identifying the 'principal Equipment' amongst other things that the *Contractor* plans to use, rather than 'the Equipment' as was stated in ECC2.

31.3 Within two weeks of the *Contractor* submitting a programme to him for acce tance, the *Project Manager* either accepts the programme or notifies the *Contr tor* of his reasons for not accepting it. A reason for not accepting a programi is that

- the *Contractor*'s plans which it shows are not practicable,
- it does not show the information which this contract requires,
- it does not represent the *Contractor*'s plans realistically or
- it does not comply with the Works Information.

Revising the programme 32

32.1 The *Contractor* shows on each revised programme

- the actual progress achieved on each operation and its effect upon t timing of the remaining work,
- the effects of implemented compensation events and of notified ea warning matters,
- how the *Contractor* plans to deal with any delays and to correct notifi Defects and
- any other changes which the *Contractor* proposes to make to the Accept Programme.

32.2 The *Contractor* submits a revised programme to the *Project Manager* for acce tance

- within the *period for reply* after the *Project Manager* has instructed him t
- when the *Contractor* chooses to and, in any case,
- at no longer interval than the interval stated in the Contract Data from t *starting date* until Completion of the whole of the *works*.

Possession of the Site 33

33.1 The *Employer* gives possession of each part of the Site to the *Contractor* on before the later of its *possession date* and the date for possession shown on t Accepted Programme.

Instructions to stop or not 34
to start work

34.1 The *Project Manager* may instruct the *Contractor* to stop or not to start a work and may later instruct him that he may re-start or start it.

Take over 35

35.1 Possession of each part of the Site returns to the *Employer* when he takes ov the part of the *works* which occupies it. Possession of the whole Site returns the *Employer* when the *Project Manager* certifies termination.

35.2 The *Employer* need not take over the *works* before the Completion Date if it stated in the Contract Data that he is not willing to do so. Otherwise t *Employer* takes over the *works* not more than two weeks after Completion.

35.3 The *Employer* may use any part of the *works* before Completion has been cer fied. If he does so, he takes over the part of the *works* when he begins to use except if the use is

- for a reason stated in the Works Information or
- to suit the *Contractor*'s method of working.

35.4 The *Project Manager* certifies the date upon which the *Employer* takes over ai part of the *works* and its extent within one week of the date.

31.3 Within two weeks of the *Contractor* submitting a programme to him for acceptance, the *Project Manager* either accepts the programme or notifies the *Contractor* of his reasons for not accepting it. A reason for not accepting a programme is that

- the *Contractor*'s plans which it shows are not practicable,
- it does not show the information which this contract requires,
- it does not represent the *Contractor*'s plans realistically or
- it does not comply with the Works Information.

Revising the programme 32

32.1 The *Contractor* shows on each revised programme

- the actual progress achieved on each operation and its effect upon the timing of the remaining work,
- the effects of implemented compensation events and of notified early warning matters,
- how the *Contractor* plans to deal with any delays and to correct notified Defects and
- any other changes which the *Contractor* proposes to make to the Accepted Programme.

32.2 The *Contractor* submits a revised programme to the *Project Manager* for acceptance

- within the *period for reply* after the *Project Manager* has instructed him to,
- when the *Contractor* chooses to and, in any case,
- at no longer interval than the interval stated in the Contract Data from the *starting date* until Completion of the whole of the *works*.

**Access to and use of 33
the Site 33.1** The *Employer* allows access to and use of each part of the Site to the *Contractor* which is necessary for the work included in this contract. Access and use is allowed on or before the later of its *access date* and the date for access shown on the Accepted Programme.

> In ECC2 the *Employer* gave 'possession' of the Site to the *Contractor* to undertake the necessary work. This is now changed to the *Employer* allowing 'access' rather than giving 'possession' and eliminates any potential problems of sites with multiple contractors working. Any constraints on access to and use of the Site should be stated in the Works Information.

**Instructions to stop or not 34
to start work 34.1** The *Project Manager* may instruct the *Contractor* to stop or not to start any work and may later instruct him that he may re-start or start it.

Take over 35

35.1 The *Employer* need not take over the *works* before the Completion Date if it is stated in the Contract Data that he is not willing to do so. Otherwise the *Employer* takes over the *works* not later than two weeks after Completion.

35.2 The *Employer* may use any part of the *works* before Completion has been certified. If he does so, he takes over the part of the *works* when he begins to use it except if the use is

- for a reason stated in the Works Information or
- to suit the *Contractor*'s method of working.

35.3 The *Project Manager* certifies the date upon which the *Employer* takes over any part of the *works* and its extent within one week of the date.

Acceleration 36

36.1   The *Project Manager* may instruct the *Contractor* to submit a quotation for acceleration to achieve Completion before the Completion Date. A quotation an acceleration comprises proposed changes to the Prices and the Completi Date and a revised programme.

36.2   The *Contractor* submits a quotation or gives his reasons for not doing so with the *period for reply*.

Acceleration 36

36.1   The *Project Manager* may instruct the *Contractor* to submit a quotation for an acceleration to achieve Completion before the Completion Date. The *Project Manager* states changes to the Key Dates to be included in the quotation. A quotation for an acceleration comprises proposed changes to the Prices and a revised programme showing the earlier Completion Date and the changed Key Dates. The *Contractor* submits details of his assessment with each quotation.

Any acceleration proposals to bring forward Completion can now include associated Key Dates. The *Contractor* is also required to submit details of his assessment with each quotation. This is something that is likely to have happened anyway under ECC2 as without such detail it would be difficult to assess.

36.2   The *Contractor* submits a quotation or gives his reasons for not doing so within the *period for reply*.

4 Testing and Defects

Tests and inspections **40**

40.1   This clause only governs tests and inspections required by the Works Informtion and the applicable law.

40.2   The *Contractor* and the *Employer* provide materials, facilities and samples f tests and inspections as stated in the Works Information.

40.3   The *Contractor* and the *Supervisor* each notifies the other of each of his te and inspections before it starts and afterwards notifies the other of its resul The *Contractor* notifies the *Supervisor* in time for a test or inspection to arranged and done before doing work which would obstruct the test or inspetion. The *Supervisor* may watch any test done by the *Contractor*.

40.4   If a test or inspection shows that any work has a Defect, the *Contractor* corre the Defect and the test or inspection is repeated.

40.5   The *Supervisor* does his tests and inspections without causing unnecessary del to the work or to a payment which is conditional upon a test or inspecti being successful. A payment which is conditional upon a *Supervisor*'s test inspection being successful becomes due at the later of the *defects date* and t end of the last *defect correction period* if

- the *Supervisor* has not done the test or inspection and
- the delay to the test or inspection is not the *Contractor*'s fault.

40.6   The *Project Manager* assesses the cost incurred by the *Employer* in repeating test or inspection after a Defect is found. The *Contractor* pays the amou assessed.

Testing and inspection **41**
before delivery 41.1   The *Contractor* does not bring to the Working Areas those Plant and Materi which the Works Information states are to be tested or inspected before delive until the *Supervisor* has notified the *Contractor* that they have passed the test inspection.

Searching and notifying **42**
Defects 42.1   The *Supervisor* may instruct the *Contractor* to search. He gives his reason f the search with his instruction. Searching may include

- uncovering, dismantling, re-covering and re-erecting work,
- providing facilities, materials and samples for tests and inspections done the *Supervisor* and
- doing tests and inspections which the Works Information does not requir

42.2   Until the *defects date*, the *Supervisor* notifies the *Contractor* of each Defe which he finds and the *Contractor* notifies the *Supervisor* of each Defect whi he finds.

Testing and Defects

Tests and inspections 40

40.1 The subclauses in this clause only apply to tests and inspections required by the Works Information or the applicable law.

40.2 The *Contractor* and the *Employer* provide materials, facilities and samples for tests and inspections as stated in the Works Information.

40.3 The *Contractor* and the *Supervisor* each notifies the other of each of his tests and inspections before it starts and afterwards notifies the other of its results. The *Contractor* notifies the *Supervisor* in time for a test or inspection to be arranged and done before doing work which would obstruct the test or inspection. The *Supervisor* may watch any test done by the *Contractor*.

40.4 If a test or inspection shows that any work has a Defect, the *Contractor* corrects the Defect and the test or inspection is repeated.

40.5 The *Supervisor* does his tests and inspections without causing unnecessary delay to the work or to a payment which is conditional upon a test or inspection being successful. A payment which is conditional upon a *Supervisor*'s test or inspection being successful becomes due at the later of the *defects date* and the end of the last *defect correction period* if

- the *Supervisor* has not done the test or inspection and
- the delay to the test or inspection is not the *Contractor*'s fault.

40.6 The *Project Manager* assesses the cost incurred by the *Employer* in repeating a test or inspection after a Defect is found. The *Contractor* pays the amount assessed.

Testing and inspection 41
before delivery

41.1 The *Contractor* does not bring to the Working Areas those Plant and Materials which the Works Information states are to be tested or inspected before delivery until the *Supervisor* has notified the *Contractor* that they have passed the test or inspection.

Searching for and 42
notifying Defects

42.1 Until the *defects date*, the *Supervisor* may instruct the *Contractor* to search for a Defect. He gives his reason for the search with his instruction. Searching may include

- uncovering, dismantling, re-covering and re-erecting work,
- providing facilities, materials and samples for tests and inspections done by the *Supervisor* and
- doing tests and inspections which the Works Information does not require.

To clarify what was generally implied in ECC2, the *Supervisor*'s powers for instructing the *Contractor* to search are in respect of a potential Defect.

42.2 Until the *defects date*, the *Supervisor* notifies the *Contractor* of each Defect as soon as he finds it and the *Contractor* notifies the *Supervisor* of each Defect as soon as he finds it.

ECC2 was silent on when each Defect requires notification. In ECC3 the requirement on both *Supervisor* and *Contractor* is to notify each other as soon as each finds a Defect.

Correcting Defects 43

43.1 The *Contractor* corrects Defects whether or not the *Supervisor* notifies him
them. The *Contractor* corrects notified Defects before the end of the *defe
correction period*. This period begins at Completion for Defects notified befo
Completion and when the Defect is notified for other Defects.

43.2 The *Supervisor* issues the Defects Certificate at the later of the *defects date* a
the end of the last *defect correction period*.

43.3 The *Project Manager* arranges for the *Employer* to give access to and use of a
part of the *works* which he has taken over to the *Contractor* if it is needed f
correcting a Defect. If the *Project Manager* has not arranged suitable access an
use within the *defect correction period*, he extends the period for correcting t
Defect as necessary.

Accepting Defects 44

44.1 The *Contractor* and the *Project Manager* may each propose to the other that t
Works Information should be changed so that a Defect does not have to l
corrected.

44.2 If the *Contractor* and the *Project Manager* are prepared to consider the chang
the *Contractor* submits a quotation for reduced Prices or an earlier Completic
Date or both to the *Project Manager* for acceptance. If the *Project Manag*
accepts the quotation, he gives an instruction to change the Works Informatio
the Prices and the Completion Date accordingly.

Uncorrected Defects 45

45.1 If the *Contractor* has not corrected a notified Defect within its *defect correctic
period*, the *Project Manager* assesses the cost of having the Defect corrected l
other people and the *Contractor* pays this amount.

Correcting Defects 43

43.1   The *Contractor* corrects a Defect whether or not the *Supervisor* notifies him of it.

43.2   The *Contractor* corrects a notified Defect before the end of the *defect correction period*. The *defect correction period* begins at Completion for Defects notified before Completion and when the Defect is notified for other Defects.

43.3   The *Supervisor* issues the Defects Certificate at the later of the *defects date* and the end of the last *defect correction period*. The *Employer*'s rights in respect of a Defect which the *Supervisor* has not found or notified are not affected by the issue of the Defects Certificate.

The additional sentence clarifies that the issue of the Defects Certificate is not conclusive evidence of absolute completion of *works* in accordance with the Works Information. Such issue also does not alter the *Contractor*'s responsibility for latent Defects or for those Defects that existed at the time of issue and not notified by either the *Supervisor* or the *Contractor*.

43.4   The *Project Manager* arranges for the *Employer* to allow the *Contractor* access to and use of a part of the *works* which he has taken over if they are needed for correcting a Defect. In this case the *defect correction period* begins when the necessary access and use have been provided.

It is sometimes not possible for an *Employer* to make available all parts of *works* taken over to suit the *Contractor*'s programme to correct Defects. This subclause is therefore a practical new inclusion that links the *defect correction period* to when the *Employer* allows access to that part of any taken over *works*.

Accepting Defects 44

44.1   The *Contractor* and the *Project Manager* may each propose to the other that the Works Information should be changed so that a Defect does not have to be corrected.

44.2   If the *Contractor* and the *Project Manager* are prepared to consider the change, the *Contractor* submits a quotation for reduced Prices or an earlier Completion Date or both to the *Project Manager* for acceptance. If the *Project Manager* accepts the quotation, he gives an instruction to change the Works Information, the Prices and the Completion Date accordingly.

Uncorrected Defects 45

45.1   If the *Contractor* is given access in order to correct a notified Defect but he has not corrected it within its *defect correction period*, the *Project Manager* assesses the cost to the *Employer* of having the Defect corrected by other people and the *Contractor* pays this amount. The Works Information is treated as having been changed to accept the Defect.

The right for the *Project Manager* to assess the cost to the *Employer* of uncorrected Defects is now linked to access. If this subclause is implemented by the *Project Manager* then the Works Information is treated as if the defective works were asked for in the first place.

45.2   If the *Contractor* is not given access in order to correct a notified Defect before the *defects date*, the *Project Manager* assesses the cost to the *Contractor* of correcting the Defect and the *Contractor* pays this amount. The Works Information is treated as having been changed to accept the Defect.

This new subclause deals with the position where the *Employer* does not give access to the *Contractor* in order to correct a Defect. The outcome is the same as subclause 45.1, where the *Project Manager* assesses the cost to the *Employer* of correcting this Defect and this amount is paid by the *Contractor*.

5 Payment

Assessing the amount due 50

50.1 The *Project Manager* assesses the amount due at each assessment date. The first assessment date is decided by the *Project Manager* to suit the procedures of the Parties and is not later than the *assessment interval* after the *starting date*. Later assessment dates occur

- at the end of each *assessment interval* until Completion of the whole of the *works*,
- at Completion of the whole of the *works*,
- four weeks after the *Supervisor* issues the Defects Certificate and
- after Completion of the whole of the *works*,
  - when an amount due is corrected and
  - when a payment is made late.

50.2 The amount due is the Price for Work Done to Date plus other amounts to be paid to the *Contractor* less amounts to be paid by or retained from the *Contractor*. Any value added tax or sales tax which the law requires the *Employer* to pay to the *Contractor* is included in the amount due.

50.3 If no programme is identified in the Contract Data, one quarter of the Price for Work Done to Date is retained in assessments of the amount due until the *Contractor* has submitted a first programme to the *Project Manager* for acceptance showing the information which this contract requires.

50.4 In assessing the amount due, the *Project Manager* considers any application for payment the *Contractor* has submitted on or before the assessment date. The *Project Manager* gives the *Contractor* details of how the amount due has been assessed.

50.5 The *Project Manager* corrects any wrongly assessed amount due in a later payment certificate.

Payment 51

51.1 The *Project Manager* certifies a payment within one week of each assessment date. The first payment is the amount due. Other payments are the change in the amount due since the last payment certificate. A payment is made by the *Contractor* to the *Employer* if the change reduces the amount due. Other payments are made by the *Employer* to the *Contractor*. Payments are in the *currency of this contract* unless otherwise stated in this contract.

51.2 Each certified payment is made within three weeks of the assessment date or, if a different period is stated in the Contract Data, within the period stated. If a payment is late, interest is paid on the late payment. Interest is assessed from the date by which the late payment should have been made until the date when the late payment is made, and is included in the first assessment after the late payment is made.

Payment

Assessing the amount due 50

50.1 The *Project Manager* assesses the amount due at each assessment date. The first assessment date is decided by the *Project Manager* to suit the procedures of the Parties and is not later than the *assessment interval* after the *starting date*. Later assessment dates occur

- at the end of each *assessment interval* until four weeks after the *Supervisor* issues the Defects Certificate and
- at Completion of the whole of the *works*.

The amendments to this subclause mean there are more assessment dates than under ECC2. There are now regular assessment dates after Completion, which reduces the potential cash flow risk to the *Contractor* that existed with the ECC2 truncated dates.

50.2 The amount due is

- the Price for Work Done to Date,
- plus other amounts to be paid to the *Contractor*,
- less amounts to be paid by or retained from the *Contractor*.

Any tax which the law requires the *Employer* to pay to the *Contractor* is included in the amount due.

50.3 If no programme is identified in the Contract Data, one quarter of the Price for Work Done to Date is retained in assessments of the amount due until the *Contractor* has submitted a first programme to the *Project Manager* for acceptance showing the information which this contract requires.

50.4 In assessing the amount due, the *Project Manager* considers any application for payment the *Contractor* has submitted on or before the assessment date. The *Project Manager* gives the *Contractor* details of how the amount due has been assessed.

50.5 The *Project Manager* corrects any wrongly assessed amount due in a later payment certificate.

Payment 51

51.1 The *Project Manager* certifies a payment within one week of each assessment date. The first payment is the amount due. Other payments are the change in the amount due since the last payment certificate. A payment is made by the *Contractor* to the *Employer* if the change reduces the amount due. Other payments are made by the *Employer* to the *Contractor*. Payments are in the *currency of this contract* unless otherwise stated in this contract.

51.2 Each certified payment is made within three weeks of the assessment date or, if a different period is stated in the Contract Data, within the period stated. If a certified payment is late, or if a payment is late because the *Project Manager* does not issue a certificate which he should issue, interest is paid on the late payment. Interest is assessed from the date by which the late payment should have been made until the date when the late payment is made, and is included in the first assessment after the late payment is made.

The second sentence has been expanded to cover certified payments being late or payments being late because the *Project Manager* does not issue a certificate he should issue. The ECC2 subclause 51.4 has been deleted in ECC3 and therefore interest is now due for one of the two stated reasons for late payment, which does not include late certification itself.

51.3 If an amount due is corrected in a later certificate either

- by the *Project Manager*, whether in relation to a mistake or a compensation event, or

- following a decision of the *Adjudicator* or the *tribunal*,

interest on the correcting amount is paid. Interest is assessed from the date when the incorrect amount was certified until the date when the correcting amount is certified and is included in the assessment which includes the correcting amount.

51.4 If the *Project Manager* does not issue a certificate which he should issue, interest is paid on the amount which he should have certified. Interest is assessed from the date by which he should have certified the amount until the date when he certifies the amount and is included in the amount then certified.

51.5 Interest is calculated at the *interest rate* and is compounded annually.

Actual Cost **52**

52.1 All the *Contractor*'s costs which are not included in the Actual Cost are deemed to be included in the *fee percentage*. Amounts included in Actual Cost are at open market or competitively tendered prices with all discounts, rebates and taxes which can be recovered deducted.

51.3   If an amount due is corrected in a later certificate either

- by the *Project Manager* in relation to a mistake or a compensation event or
- following a decision of the *Adjudicator* or the *tribunal*,

interest on the correcting amount is paid. Interest is assessed from the date when the incorrect amount was certified until the date when the correcting amount is certified and is included in the assessment which includes the correcting amount.

51.4   Interest is calculated on a daily basis at the *interest rate* and is compounded annually.

The slight amendment to this subclause results in clarifying the basis for calculating interest due, that being on a daily basis.

Defined Cost   52

52.1   All the *Contractor*'s costs which are not included in the Defined Cost are treated as included in the Fee. Defined Cost includes only amounts calculated using rates and percentages stated in the Contract Data and other amounts at open market or competitively tendered prices with deductions for all discounts, rebates and taxes which can be recovered.

'Actual Cost' has been replaced with 'Defined Cost'. This is a very practical amendment as users commonly thought of 'Actual Cost' as being real actual cost whereas it never was. Each main Option defines precisely what is meant by the term 'Defined Cost'.

6 Compensation events

Compensation events **60**

60.1 The following are compensation events.

(1) The *Project Manager* gives an instruction changing the Works Information except

- a change made in order to accept a Defect or
- a change to the Works Information provided by the *Contractor* for his design which is made at his request or to comply with other Works Information provided by the *Employer*.

(2) The *Employer* does not give possession of a part of the Site by the later of its *possession date* and the date required by the Accepted Programme.

(3) The *Employer* does not provide something which he is to provide by the date for providing it required by the Accepted Programme.

(4) The *Project Manager* gives an instruction to stop or not to start any work.

(5) The *Employer* or Others do not work within the times shown on the Accepted Programme or do not work within the conditions stated in the Works Information.

(6) The *Project Manager* or the *Supervisor* does not reply to a communication from the *Contractor* within the period required by this contract.

(7) The *Project Manager* gives an instruction for dealing with an object of value or of historical or other interest found within the Site.

(8) The *Project Manager* or the *Supervisor* changes a decision which he has previously communicated to the *Contractor*.

(9) The *Project Manager* withholds an acceptance (other than acceptance of a quotation for acceleration or for not correcting a Defect) for a reason not stated in this contract.

(10) The *Supervisor* instructs the *Contractor* to search and no Defect is found unless the search is needed only because the *Contractor* gave insufficient notice of doing work obstructing a required test or inspection.

(11) A test or inspection done by the *Supervisor* causes unnecessary delay.

Compensation events

Compensation events   60

60.1   The following are compensation events.

(1) The *Project Manager* gives an instruction changing the Works Information except

- a change made in order to accept a Defect or
- a change to the Works Information provided by the *Contractor* for his design which is made either at his request or to comply with other Works Information provided by the *Employer*.

(2) The *Employer* does not allow access to and use of a part of the Site by the later of its *access date* and the date shown on the Accepted Programme.

Amendments to this subclause line up with the change from 'possession' to 'access' elsewhere.

(3) The *Employer* does not provide something which he is to provide by the date for providing it shown on the Accepted Programme.

A minor but practical change replacing 'required by the Accepted Programme' with 'shown on the Accepted Programme'. The something to be provided by the *Employer* should be stated in the Works Information which in turn should be shown on the programme.

(4) The *Project Manager* gives an instruction to stop or not to start any work or to change a Key Date.

(5) The *Employer* or Others

- do not work within the times shown on the Accepted Programme,
- do not work within the conditions stated in the Works Information or
- carry out work on the Site that is not stated in the Works Information.

This subclause is now bulleted for certain stated defaults by the *Employer* or Others and now has a third event provided which covers works being carried out on Site by the *Employer* or Others that were not stated in the Works Information. This emphasises the care that is required in accurately drafting the Works Information at time of tender.

(6) The *Project Manager* or the *Supervisor* does not reply to a communication from the *Contractor* within the period required by this contract.

(7) The *Project Manager* gives an instruction for dealing with an object of value or of historical or other interest found within the Site.

(8) The *Project Manager* or the *Supervisor* changes a decision which he has previously communicated to the *Contractor*.

(9) The *Project Manager* withholds an acceptance (other than acceptance of a quotation for acceleration or for not correcting a Defect) for a reason not stated in this contract.

(10) The *Supervisor* instructs the *Contractor* to search for a Defect and no Defect is found unless the search is needed only because the *Contractor* gave insufficient notice of doing work obstructing a required test or inspection.

(11) A test or inspection done by the *Supervisor* causes unnecessary delay.

(12) The *Contractor* encounters physical conditions which

- are within the Site,
- are not weather conditions and
- which an experienced contractor would have judged at the Contract Date to have such a small chance of occurring that it would have been unreasonable for him to have allowed for them.

(13) A *weather measurement* is recorded

- within a calendar month,
- before the Completion Date for the whole of the *works* and
- at the place stated in the Contract Data

the value of which, by comparison with the *weather data*, is shown to occur on average less frequently than once in ten years.

(14) An *Employer*'s risk event occurs.

(15) The *Project Manager* certifies take over of a part of the *works* before both Completion and the Completion Date.

(16) The *Employer* does not provide materials, facilities and samples for tests as stated in the Works Information.

(17) The *Project Manager* notifies a correction to an assumption about the nature of a compensation event.

(18) A breach of contract by the *Employer* which is not one of the other compensation events in this contract.

(12) The *Contractor* encounters physical conditions which

- are within the Site,
- are not weather conditions and
- an experienced contractor would have judged at the Contract Date to have such a small chance of occurring that it would have been unreasonable for him to have allowed for them.

Only the difference between the physical conditions encountered and those for which it would have been reasonable to have allowed is taken into account in assessing a compensation event.

The additional paragraph to this subclause confirms the intention of ECC2 that it is only the difference between the physical conditions encountered and those it would have been reasonable to have allowed for is the basis of assessing this compensation event.

(13) A *weather measurement* is recorded

- within a calendar month,
- before the Completion Date for the whole of the *works* and
- at the place stated in the Contract Data

the value of which, by comparison with the *weather data*, is shown to occur on average less frequently than once in ten years.

Only the difference between the *weather measurement* and the weather which the *weather data* show to occur on average less frequently than once in ten years is taken into account in assessing a compensation event.

In a similar approach to subclause 60.1(12), the additional paragraph to this subclause also confirms that it is only the difference between the *weather measurement* and the weather which *weather data* shows to occur on average less frequently than once in ten years that is the basis of assessing this compensation event.

(14) An event which is an *Employer*'s risk stated in this contract.

The changes to this subclause clarify that such a compensation event arises where the event is stated as an *Employer*'s risk in this contract, which means those in subclause 80.1 together with any additional *Employer*'s risks stated in the Contract Data.

(15) The *Project Manager* certifies take over of a part of the *works* before both Completion and the Completion Date.

(16) The *Employer* does not provide materials, facilities and samples for tests and inspections as stated in the Works Information.

(17) The *Project Manager* notifies a correction to an assumption which he has stated about a compensation event.

(18) A breach of contract by the *Employer* which is not one of the other compensation events in this contract.

(19) An event which

- stops the *Contractor* completing the *works* or
- stops the *Contractor* completing the *works* by the date shown on the Accepted Programme,

60.2   In judging the physical conditions, the *Contractor* is assumed to have taken into account

- the Site Information,
- publicly available information referred to in the Site Information,
- information obtainable from a visual inspection of the Site and
- other information which an experienced contractor could reasonably be expected to have or to obtain.

60.3   If there is an inconsistency within the Site Information (including the information referred to in it), the *Contractor* is assumed to have taken into account the physical conditions more favourable to doing the work.

Notifying compensation events   **61**

61.1   For compensation events which arise from the *Project Manager* or the *Supervisor* giving an instruction or changing an earlier decision, the *Project Manager* notifies the *Contractor* of the compensation event at the time of the event. He also instructs the *Contractor* to submit quotations, unless the event arises from a fault of the *Contractor* or quotations have already been submitted. The *Contractor* puts the instruction or changed decision into effect.

61.2   The *Project Manager* may instruct the *Contractor* to submit quotations for a proposed instruction or a proposed changed decision. The *Contractor* does not put a proposed instruction or a proposed changed decision into effect.

61.3   The *Contractor* notifies an event which has happened or which he expects to happen to the *Project Manager* as a compensation event if

- the *Contractor* believes that the event is a compensation event,
- it is less than two weeks since he became aware of the event and
- the *Project Manager* has not notified the event to the *Contractor*.

and which

- neither Party could prevent,
- an experienced contractor would have judged at the Contract Date to have such a small chance of occurring that it would have been unreasonable for him to have allowed for it and
- is not one of the other compensation events stated in this contract.

This new compensation event provides the *Contractor* with relief for those events described as 'prevention' in subclause 19.1. As with all other compensation events, changes to the Prices, any Key Dates and the Completion Date are assessed.

60.2   In judging the physical conditions for the purpose of assessing a compensation event, the *Contractor* is assumed to have taken into account

- the Site Information,
- publicly available information referred to in the Site Information,
- information obtainable from a visual inspection of the Site and
- other information which an experienced contractor could reasonably be expected to have or to obtain.

60.3   If there is an ambiguity or inconsistency within the Site Information (including the information referred to in it), the *Contractor* is assumed to have taken into account the physical conditions more favourable to doing the work.

In terms of using the Site Information for judging the physical conditions under subclauses 60.1(12) and 60.2, subclause 60.3 now includes a basis for interpreting ambiguities within the Site Information as well as inconsistencies.

Notifying compensation **61**
events 61.1   For compensation events which arise from the *Project Manager* or the *Supervisor* giving an instruction or changing an earlier decision, the *Project Manager* notifies the *Contractor* of the compensation event at the time of giving the instruction or changing the earlier decision. He also instructs the *Contractor* to submit quotations, unless the event arises from a fault of the *Contractor* or quotations have already been submitted. The *Contractor* puts the instruction or changed decision into effect.

61.2   The *Project Manager* may instruct the *Contractor* to submit quotations for a proposed instruction or a proposed changed decision. The *Contractor* does not put a proposed instruction or a proposed changed decision into effect.

61.3   The *Contractor* notifies the *Project Manager* of an event which has happened or which he expects to happen as a compensation event if

- the *Contractor* believes that the event is a compensation event and
- the *Project Manager* has not notified the event to the *Contractor*.

If the *Contractor* does not notify a compensation event within eight weeks of becoming aware of the event, he is not entitled to a change in the Prices, the Completion Date or a Key Date unless the *Project Manager* should have notified the event to the *Contractor* but did not.

In ECC2 the *Contractor* was obliged to notify a compensation event within two weeks of becoming aware of it. There was no clear provision for what happened if the time limit was missed. A significant change in ECC3 of particular interest to the *Contractor* is the addition of a new paragraph extending the time period for notifying the compensation event to eight weeks from being aware of the event. It goes on to state that the *Contractor* is not otherwise entitled to a change in the Prices, the Completion Date or a Key Date. This new provision only applies to those events that the *Project Manager* is not obliged to notify so, for example, the common occurrence of the *Project Manager* giving instructions to change the Works Information is not caught by this strict limitation. To ensure he suffers no financial hardship, the *Contractor* needs to ensure he has an effective process for managing compensation events.

61.4 The Prices and the Completion Date are not changed if the *Project Manager* decides that an event notified by the *Contractor*

- arises from a fault of the *Contractor*,
- has not happened and is not expected to happen,
- has no effect upon Actual Cost or Completion or
- is not one of the compensation events stated in this contract.

If the *Project Manager* decides otherwise, he instructs the *Contractor* to submit quotations for the event. Within either

- one week of the *Contractor*'s notification or
- a longer period to which the *Contractor* has agreed

the *Project Manager* notifies his decision to the *Contractor* or instructs him to submit quotations.

61.5 If the *Project Manager* decides that the *Contractor* did not give an early warning of the event which an experienced contractor could have given, he notifies this decision to the *Contractor* when he instructs him to submit quotations.

61.6 If the *Project Manager* decides that the effects of a compensation event are too uncertain to be forecast reasonably, he states assumptions about the event in his instruction to the *Contractor* to submit quotations. Assessment of the event is based on these assumptions. If any of them is later found to have been wrong, the *Project Manager* notifies a correction .

61.7 A compensation event is not notified after the *defects date*.

Quotations for **62**
compensation events 62.1 The *Project Manager* may instruct the *Contractor* to submit alternative quotations based upon different ways of dealing with the compensation event which are practicable. The *Contractor* submits the required quotations to the *Project Manager* and may submit quotations for other methods of dealing with the compensation event which he considers practicable.

61.4   If the *Project Manager* decides that an event notified by the *Contractor*

- arises from a fault of the *Contractor*,
- has not happened and is not expected to happen,
- has no effect upon Defined Cost, Completion or meeting a Key Date or
- is not one of the compensation events stated in this contract

he notifies the *Contractor* of his decision that the Prices, the Completion Date and the Key Dates are not to be changed.

If the *Project Manager* decides otherwise, he notifies the *Contractor* accordingly and instructs him to submit quotations.

If the *Project Manager* does not notify his decision to the *Contractor* within either

- one week of the *Contractor*'s notification or
- a longer period to which the *Contractor* has agreed,

the *Contractor* may notify the *Project Manager* to this effect. A failure by the *Project Manager* to reply within two weeks of this notification is treated as acceptance by the *Project Manager* that the event is a compensation event and an instruction to submit quotations.

The amendments to this subclause are part of a three-part attack on the so called 'indolent *Project Manager*'. Experience has shown that occasionally a *Project Manager*, despite having numerous options available, is not keen to make decisions in terms of whether an event is indeed a compensation event or maybe whether the quotation seems to have been assessed correctly by the *Contractor*. The preference is sometimes to revert to using hindsight rather than foresight, which is considered to be less advantageous to the Parties. The new provision is really a reminder to act in accordance with the contract, failure to act may result in the *Employer* being disadvantaged as silence from the *Project Manager* results in deemed acceptance that the event is a compensation event and an instruction to submit quotations. The *Project Manager* potentially gets a notification and a further reminder to make a decision under this subclause. This should be enough of a reminder for any competent person to act.

61.5   If the *Project Manager* decides that the *Contractor* did not give an early warning of the event which an experienced contractor could have given, he notifies this decision to the *Contractor* when he instructs him to submit quotations.

61.6   If the *Project Manager* decides that the effects of a compensation event are too uncertain to be forecast reasonably, he states assumptions about the event in his instruction to the *Contractor* to submit quotations. Assessment of the event is based on these assumptions. If any of them is later found to have been wrong, the *Project Manager* notifies a correction.

61.7   A compensation event is not notified after the *defects date*.

Quotations for   **62**
compensation events   62.1   After discussing with the *Contractor* different ways of dealing with the compensation event which are practicable, the *Project Manager* may instruct the *Contractor* to submit alternative quotations. The *Contractor* submits the required quotations to the *Project Manager* and may submit quotations for other methods of dealing with the compensation event which he considers practicable.

The changes in this subclause reinforce the emphasis on getting the *Project Manager* and *Contractor* to jointly engage in looking at different ways of dealing with each compensation event. It is about collaborative working, not a master and servant approach.

62.2 Quotations for compensation events comprise proposed changes to the Prices and any delay to the Completion Date assessed by the *Contractor*. The *Contractor* submits details of his assessment with each quotation. If the programme for remaining work is affected by the compensation event, the *Contractor* includes a revised programme in his quotation showing the effect.

62.3 The *Contractor* submits quotations within three weeks of being instructed to do so by the *Project Manager*. The *Project Manager* replies within two weeks of the submission. His reply is

- an instruction to submit a revised quotation,

- an acceptance of a quotation,

- a notification that a proposed instruction or a proposed changed decision will not be given or

- a notification that he will be making his own assessment.

62.4 The *Project Manager* instructs the *Contractor* to submit a revised quotation only after explaining his reasons for doing so to the *Contractor*. The *Contractor* submits the revised quotation within three weeks of being instructed to do so.

62.5 The *Project Manager* extends the time allowed for

- the *Contractor* to submit quotations for a compensation event and

- the *Project Manager* to reply to a quotation

if the *Project Manager* and the *Contractor* agree to the extension before the submission or reply is due. The *Project Manager* notifies the extension that has been agreed to the *Contractor*.

62.2   Quotations for compensation events comprise proposed changes to the Prices and any delay to the Completion Date and Key Dates assessed by the *Contractor*. The *Contractor* submits details of his assessment with each quotation. If the programme for remaining work is altered by the compensation event, the *Contractor* includes the alterations to the Accepted Programme in his quotation.

This subclause contains a small but significant change to submitting revised programmes. In ECC2 the *Contractor* was required to submit a revised programme with each quotation for each compensation event, which resulted in potentially many revised programmes being submitted. In ECC3 the *Contractor* is obliged to include the alterations to the Accepted Programme in his programme. This should help to more easily identify the effects of a compensation event and reduce the number of entire revised programmes submitted. The Completion Date can still be revised, though, on the basis of the Accepted Programme together with the accepted quotation altering this date.

62.3   The *Contractor* submits quotations within three weeks of being instructed to do so by the *Project Manager*. The *Project Manager* replies within two weeks of the submission. His reply is

- an instruction to submit a revised quotation,
- an acceptance of a quotation,
- a notification that a proposed instruction will not be given or a proposed changed decision will not be made or
- a notification that he will be making his own assessment.

62.4   The *Project Manager* instructs the *Contractor* to submit a revised quotation only after explaining his reasons for doing so to the *Contractor*. The *Contractor* submits the revised quotation within three weeks of being instructed to do so.

62.5   The *Project Manager* extends the time allowed for

- the *Contractor* to submit quotations for a compensation event and
- the *Project Manager* to reply to a quotation

if the *Project Manager* and the *Contractor* agree to the extension before the submission or reply is due. The *Project Manager* notifies the extension that has been agreed to the *Contractor*.

62.6   If the *Project Manager* does not reply to a quotation within the time allowed, the *Contractor* may notify the *Project Manager* to this effect. If the *Contractor* submitted more than one quotation for the compensation event, he states in his notification which quotation he proposes is to be accepted. If the *Project Manager* does not reply to the notification within two weeks, and unless the quotation is for a proposed instruction or a proposed changed decision, the *Contractor*'s notification is treated as acceptance of the quotation by the *Project Manager*.

This is the second part of the attack on the indolent *Project Manager*. Once again there is deemed acceptance, but this time of a *Contractor*'s quotation where the *Project Manager* has failed to reply to a quotation both within the time allowed and following a further reminder from the *Contractor*. If the *Contractor* does not so remind and the *Project Manager* replies to the quotation later than allowed, then a compensation event will arise for late reply under subclause 60.1(6).

**Assessing compensation 63
events** 63.1 The changes to the Prices are assessed as the effect of the compensation event upon

- the Actual Cost of the work already done,
- the forecast Actual Cost of the work not yet done and
- the resulting Fee.

63.2 If the effect of a compensation event is to reduce the total Actual Cost, the Prices are not reduced except as stated in this contract.

63.3 A delay to the Completion Date is assessed as the length of time that, due to the compensation event, planned Completion is later than planned Completion as shown on the Accepted Programme.

63.4 If the *Project Manager* has notified the *Contractor* of his decision that the *Contractor* did not give an early warning of a compensation event which an experienced contractor could have given, the event is assessed as if the *Contractor* had given early warning.

63.5 Assessment of the effect of a compensation event includes cost and time risk allowances for matters which have a significant chance of occurring and are at the *Contractor*'s risk under this contract.

63.6 Assessments are based upon the assumptions that the *Contractor* reacts competently and promptly to the compensation event, that the additional Actual Cost and time due to the event are reasonably incurred and that the Accepted Programme can be changed.

Assessing compensation   **63**

events   63.1   The changes to the Prices are assessed as the effect of the compensation event upon

- the actual Defined Cost of the work already done,
- the forecast Defined Cost of the work not yet done and
- the resulting Fee.

The date when the *Project Manager* instructed or should have instructed the *Contractor* to submit quotations divides the work already done from the work not yet done.

> In ECC2 the split between assessing Actual Cost and forecast Actual Cost in a compensation event was not clear. Was it when the event occurred, when the quotation was instructed or when the quotation was provided? Apart from amending 'Actual Cost' to 'actual Defined Cost', the additional provisions in this subclause refer to when the *Project Manager* did instruct or should have instructed the *Contractor* to submit quotations as being the dividing date for assessing such a compensation event.

63.2   If the effect of a compensation event is to reduce the total Defined Cost, the Prices are not reduced except as stated in this contract.

> The remaining parts of ECC2 subclause 63.2 have been moved to subclause 63.10 in main Options A and B, and 63.11 in main Options C and D.

63.3   A delay to the Completion Date is assessed as the length of time that, due to the compensation event, planned Completion is later than planned Completion as shown on the Accepted Programme. A delay to a Key Date is assessed as the length of time that, due to the compensation event, the planned date when the Condition stated for a Key Date will be met is later than the date shown on the Accepted Programme.

> The basis for assessing delays to a Key Date is similar to that for assessing delays to planned Completion. Using the Accepted Programme, one has regard to the effect of the compensation event upon the date when the *Contractor* plans to meet the Condition for a Key Date.

63.4   The rights of the *Employer* and the *Contractor* to changes to the Prices, the Completion Date and the Key Dates are their only rights in respect of a compensation event.

> A new subclause stating that in terms of a compensation event, the Parties' only rights are to change the Prices, the Completion Date and a Key Date.

63.5   If the *Project Manager* has notified the *Contractor* of his decision that the *Contractor* did not give an early warning of a compensation event which an experienced contractor could have given, the event is assessed as if the *Contractor* had given early warning.

63.6   Assessment of the effect of a compensation event includes risk allowances for cost and time for matters which have a significant chance of occurring and are at the *Contractor*'s risk under this contract.

> Instead of the ECC2 provision of 'cost and time risk allowances' the change to 'risk allowances for cost and time' gives a clearer description of what this is supposed to cover.

63.7 A compensation event which is an instruction to change the Works Information in order to resolve an ambiguity or inconsistency is assessed as follows. If Works Information provided by the *Employer* is changed, the effect of the compensation event is assessed as if the Prices and the Completion Date were for the interpretation most favourable to the *Contractor*. If Works Information provided by the *Contractor* is changed, the effect of the compensation event is assessed as if the Prices and the Completion Date were for the interpretation most favourable to the *Employer*.

The *Project Manager*'s 64
Assessments 64.1 The *Project Manager* assesses a compensation event

- if the *Contractor* has not submitted a required quotation and details of his assessment within the time allowed,

- if the *Project Manager* decides that the *Contractor* has not assessed the compensation event correctly in a quotation and he does not instruct the *Contractor* to submit a revised quotation,

- if, when the *Contractor* submits quotations for a compensation event, he has not submitted a programme which this contract requires him to submit or

- if when the *Contractor* submits quotations for a compensation event the *Project Manager* has not accepted the *Contractor*'s latest programme for one of the reasons stated in this contract.

64.2 The *Project Manager* assesses a compensation event using his own assessment of the programme for the remaining work if

- there is no Accepted Programme or

- the *Contractor* has not submitted a revised programme for acceptance as required by this contract.

64.3 The *Project Manager* notifies the *Contractor* of his assessment of a compensation event and gives him details of it within the period allowed for the *Contractor*'s submission of his quotation for the same event. This period starts when the need for the *Project Manager*'s assessment becomes apparent.

63.7 Assessments are based upon the assumptions that the *Contractor* reacts competently and promptly to the compensation event, that any Defined Cost and time due to the event are reasonably incurred and that the Accepted Programme can be changed.

63.8 A compensation event which is an instruction to change the Works Information in order to resolve an ambiguity or inconsistency is assessed as if the Prices, the Completion Date and the Key Dates were for the interpretation most favourable to the Party which did not provide the Works Information.

> This subclause has been re-drafted to shorten the wording but not change the intent, that is ambiguities or inconsistencies are resolved in favour of the Party which did not provide the Works Information.

63.9 If a change to the Works Information makes the description of the Condition for a Key Date incorrect, the *Project Manager* corrects the description. This correction is taken into account in assessing the compensation event for the change to the Works Information.

The *Project Manager*'s assessments

64

64.1 The *Project Manager* assesses a compensation event

- if the *Contractor* has not submitted a quotation and details of his assessment within the time allowed,
- if the *Project Manager* decides that the *Contractor* has not assessed the compensation event correctly in a quotation and he does not instruct the *Contractor* to submit a revised quotation,
- if, when the *Contractor* submits quotations for a compensation event, he has not submitted a programme or alterations to a programme which this contract requires him to submit or
- if, when the *Contractor* submits quotations for a compensation event, the *Project Manager* has not accepted the *Contractor*'s latest programme for one of the reasons stated in this contract.

64.2 The *Project Manager* assesses a compensation event using his own assessment of the programme for the remaining work if

- there is no Accepted Programme or
- the *Contractor* has not submitted a programme or alterations to a programme for acceptance as required by this contract.

64.3 The *Project Manager* notifies the *Contractor* of his assessment of a compensation event and gives him details of it within the period allowed for the *Contractor*'s submission of his quotation for the same event. This period starts when the need for the *Project Manager*'s assessment becomes apparent.

64.4 If the *Project Manager* does not assess a compensation event within the time allowed, the *Contractor* may notify the *Project Manager* to this effect. If the *Contractor* submitted more than one quotation for the compensation event, he states in his notification which quotation he proposes is to be accepted. If the *Project Manager* does not reply within two weeks of this notification the notification is treated as acceptance of the *Contractor*'s quotation by the *Project Manager*.

> This is the third part of the attack on the indolent *Project Manager*. Once again there is deemed acceptance, but this time of a *Contractor*'s quotation where the *Project Manager* has failed to assess a compensation event both within the time allowed and following a further reminder from the *Contractor*. If the *Contractor* does not so remind and the *Project Manager* assesses the compensation event later than allowed, then a compensation event will arise for late reply under subclause 60.1(6). The *Project Manager* usually only invokes clause 64 where there has been some sort of failure by the *Contractor* in the assessment process. For a deemed acceptance of a quotation to apply under subclause 64.4, there must be a quotation submitted by the *Contractor*.

Implementing compensation **65**
events

65.1 The *Project Manager* implements each compensation event by notifying the *Contractor* of the quotation which he has accepted or of his own assessment. He implements the compensation event when he accepts a quotation or completes his own assessment or when the compensation event occurs, whichever is latest.

65.2 The assessment of a compensation event is not revised if a forecast upon which it is based is shown by later recorded information to have been wrong.

Implementing compensation **65**

Implementing **65**
compensation events 65.1   A compensation event is implemented when

- the *Project Manager* notifies his acceptance of the *Contractor*'s quotation,
- the *Project Manager* notifies the *Contractor* of his own assessment or
- a *Contractor*'s quotation is treated as having been accepted by the *Project Manager*.

65.2   The assessment of a compensation event is not revised if a forecast upon which it is based is shown by later recorded information to have been wrong.

7 Title

The *Employer*'s title to **70**
Equipment, Plant and 70.1 Whatever title the *Contractor* has to Equipment, Plant and Materials which is
Materials outside the Working Areas passes to the *Employer* if the *Supervisor* has marked
it as for this contract.

70.2 Whatever title the *Contractor* has to Equipment, Plant and Materials passes to
the *Employer* if it has been brought within the Working Areas. The title to
Equipment, Plant and Materials passes back to the *Contractor* if it is removed
from the Working Areas with the *Project Manager*'s permission.

Marking Equipment, Plant **71**
and Materials outside the 71.1 The *Supervisor* marks Equipment, Plant and Materials which are outside the
Working Areas Working Areas if

- this contract identifies them for payment and

- the *Contractor* has prepared them for marking as the Works Information
requires.

Removing Equipment **72**
72.1 The *Contractor* removes Equipment from the Site when it is no longer needed
unless the *Project Manager* allows it to be left in the *works*.

Objects and materials **73**
within the Site 73.1 The *Contractor* has no title to an object of value or of historical or other interest
within the Site. The *Contractor* notifies the *Project Manager* when such an
object is found and the *Project Manager* instructs the *Contractor* how to deal
with it. The *Contractor* does not move the object without instructions.

73.2 The *Contractor* has title to materials from excavation and demolition only as
stated in the Works Information.

Title

The *Employer*'s title to Plant and Materials	**70**	
	70.1	Whatever title the *Contractor* has to Plant and Materials which is outside the Working Areas passes to the *Employer* if the *Supervisor* has marked it as for this contract.
	70.2	Whatever title the *Contractor* has to Plant and Materials passes to the *Employer* if it has been brought within the Working Areas. The title to Plant and Materials passes back to the *Contractor* if it is removed from the Working Areas with the *Project Manager*'s permission.

References to 'Equipment' have been taken out of clause 70. It was not intended that the *Employer* has title to the likes of the *Contractor*'s accommodation or the mechanical plant he is constructing the *works* with until it is removed from the Working Areas.

Marking Equipment, Plant and Materials outside the Working Areas	**71**	
	71.1	The *Supervisor* marks Equipment, Plant and Materials which are outside the Working Areas if

- this contract identifies them for payment and
- the *Contractor* has prepared them for marking as the Works Information requires.

Removing Equipment	**72**	
	72.1	The *Contractor* removes Equipment from the Site when it is no longer needed unless the *Project Manager* allows it to be left in the *works*.

Objects and materials within the Site	**73**	
	73.1	The *Contractor* has no title to an object of value or of historical or other interest within the Site. The *Contractor* notifies the *Project Manager* when such an object is found and the *Project Manager* instructs the *Contractor* how to deal with it. The *Contractor* does not move the object without instructions.
	73.2	The *Contractor* has title to materials from excavation and demolition only as stated in the Works Information.

8 Risks and insurance

Employer's risks **80**

80.1 The *Employer*'s risks are

- Claims, proceedings, compensation and costs payable which are due to

  - use or occupation of the Site by the *works* or for the purpose of the *works* which is the unavoidable result of the *works*,

  - negligence, breach of statutory duty or interference with any legal right by the *Employer* or by any person employed by or contracted to him except the *Contractor* or

  - a fault of the *Employer* or a fault in his design.

- Loss of or damage to Plant and Materials supplied to the *Contractor* by the *Employer*, or by Others on the *Employer*'s behalf, until the *Contractor* has received and accepted them.

- Loss of or damage to the *works*, Plant and Materials due to

  - war, civil war, rebellion, revolution, insurrection, military or usurped power,

  - strikes, riots and civil commotion not confined to the *Contractor*'s employees,

  - radioactive contamination.

- Loss of or damage to the parts of the *works* taken over by the *Employer*, except loss or damage occurring before the issue of the Defects Certificate which is due to

  - a Defect which existed at take over,

  - an event occurring before take over which was not itself an *Employer*'s risk or

  - the activities of the *Contractor* on the Site after take over.

- Loss of or damage to the *works* and any Equipment, Plant and Materials retained on the Site by the *Employer* after a termination, except loss and damage due to the activities of the *Contractor* on the Site after the termination.

- Additional *Employer*'s risks stated in the Contract Data.

The *Contractor*'s risks **81**

81.1 From the *starting date* until the Defects Certificate has been issued the risk which are not carried by the *Employer* are carried by the *Contractor*.

Repairs **82**

82.1 Until the Defects Certificate has been issued and unless otherwise instructed by the *Project Manager* the *Contractor* promptly replaces loss of and repair damage to the *works*, Plant and Materials.

Indemnity **83**

83.1 Each Party indemnifies the other against claims, proceedings, compensation and costs due to an event which is at his risk.

83.2 The liability of each Party to indemnify the other is reduced if events at the other Party's risk contributed to the claims, proceedings, compensation and costs. The reduction is in proportion to the extent that events which were at the other Party's risk contributed, taking into account each Party's responsibilities under this contract.

Risks and insurance

Employer's risks   **80**

80.1   The following are *Employer*'s risks.

- Claims, proceedings, compensation and costs payable which are due to

  - use or occupation of the Site by the *works* or for the purpose of the *works* which is the unavoidable result of the *works*,
  - negligence, breach of statutory duty or interference with any legal right by the *Employer* or by any person employed by or contracted to him except the *Contractor* or
  - a fault of the *Employer* or a fault in his design.

- Loss of or damage to Plant and Materials supplied to the *Contractor* by the *Employer*, or by Others on the *Employer*'s behalf, until the *Contractor* has received and accepted them.

- Loss of or damage to the *works*, Plant and Materials due to

  - war, civil war, rebellion, revolution, insurrection, military or usurped power,
  - strikes, riots and civil commotion not confined to the *Contractor*'s employees or
  - radioactive contamination.

- Loss of or wear or damage to the parts of the *works* taken over by the *Employer*, except loss, wear or damage occurring before the issue of the Defects Certificate which is due to

  - a Defect which existed at take over,
  - an event occurring before take over which was not itself an *Employer*'s risk or
  - the activities of the *Contractor* on the Site after take over.

- Loss of or wear or damage to the *works* and any Equipment, Plant and Materials retained on the Site by the *Employer* after a termination, except loss, wear or damage due to the activities of the *Contractor* on the Site after the termination.

- Additional *Employer*'s risks stated in the Contract Data.

'Wear' has been added to certain of the stated *Employer*'s risks, which is a practical inclusion.

The *Contractor*'s risks   **81**

81.1   From the *starting date* until the Defects Certificate has been issued, the risks which are not carried by the *Employer* are carried by the *Contractor*.

Repairs   **82**

82.1   Until the Defects Certificate has been issued and unless otherwise instructed by the *Project Manager*, the *Contractor* promptly replaces loss of and repairs damage to the *works*, Plant and Materials.

Indemnity   **83**

83.1   Each Party indemnifies the other against claims, proceedings, compensation and costs due to an event which is at his risk.

83.2   The liability of each Party to indemnify the other is reduced if events at the other Party's risk contributed to the claims, proceedings, compensation and costs. The reduction is in proportion to the extent that events which were at the other Party's risk contributed, taking into account each Party's responsibilities under this contract.

Insurance cover **84**

84.1   The *Contractor* provides the insurances stated in the Insurance Table except any insurance which the *Employer* is to provide as stated in the Contract Data. The *Contractor* provides additional insurances as stated in the Contract Data.

84.2   The insurances are in the joint names of the Parties and provide cover for events which are at the *Contractor*'s risk from the *starting date* until the Defects Certificate has been issued.

INSURANCE TABLE

Insurance against	Minimum amount of cover or minimum limit of indemnity
Loss of or damage to the *works*, Plant and Materials.	The replacement cost, including the amount stated in the Contract Data for the replacement of any Plant and Materials provided by the *Employer*.
Loss of or damage to Equipment.	The replacement cost.
Liability for loss of or damage to property (except the *works*, Plant and Materials and Equipment) and liability for bodily injury to or death of a person (not an employee of the *Contractor*) caused by activity in connection with this contract.	The amount stated in the Contract Data for any one event with cross liability so that the insurance applies to the Parties separately.
Liability for death of or bodily injury to employees of the *Contractor* arising out of and in the course of their employment in connection with this contract.	The greater of the amount required by the applicable law and the amount stated in the Contract Data for any one event.

Insurance policies **85**

85.1   The *Contractor* submits policies and certificates for the insurance which he is to provide to the *Project Manager* for acceptance before the *starting date* and afterwards as the *Project Manager* instructs. A reason for not accepting the policies and certificates is that they do not comply with this contract.

85.2   Insurance policies include a waiver by the insurers of their subrogation right against directors and other employees of every insured except where there i fraud.

85.3   The Parties comply with the terms and conditions of the insurance policies.

85.4   Any amount not recovered from an insurer is borne by the *Employer* for event which are at his risk and by the *Contractor* for events which are at his risk.

If the *Contractor* does not **86**
insure 86.1   The *Employer* may insure a risk which this contract requires the *Contractor* t insure if the *Contractor* does not submit a required policy or certificate. The co of this insurance to the *Employer* is paid by the *Contractor*.

Insurance cover **84**

84.1 The *Contractor* provides the insurances stated in the Insurance Table except any insurance which the *Employer* is to provide as stated in the Contract Data. The *Contractor* provides additional insurances as stated in the Contract Data.

84.2 The insurances are in the joint names of the Parties and provide cover for events which are at the *Contractor*'s risk from the *starting date* until the Defects Certificate or a termination certificate has been issued.

INSURANCE TABLE

Insurance against	Minimum amount of cover or minimum limit of indemnity
Loss of or damage to the *works*, Plant and Materials	The replacement cost, including the amount stated in the Contract Data for the replacement of any Plant and Materials provided by the *Employer*
Loss of or damage to Equipment	The replacement cost
Liability for loss of or damage to property (except the *works*, Plant and Materials and Equipment) and liability for bodily injury to or death of a person (not an employee of the *Contractor*) caused by activity in connection with this contract	The amount stated in the Contract Data for any one event with cross liability so that the insurance applies to the Parties separately
Liability for death of or bodily injury to employees of the *Contractor* arising out of and in the course of their employment in connection with this contract	The greater of the amount required by the applicable law and the amount stated in the Contract Data for any one event

Insurance policies **85**

85.1 Before the *starting date* and on each renewal of the insurance policy until the *defects date*, the *Contractor* submits to the *Project Manager* for acceptance certificates which state that the insurance required by this contract is in force. The certificates are signed by the *Contractor*'s insurer or insurance broker. A reason for not accepting the certificates is that they do not comply with this contract.

Instead of submitting insurance policies, as required by ECC2, the *Contractor* now submits certificates which are signed by the *Contractor*'s insurer or insurance broker and state that the insurances required by the contract are in force.

85.2 Insurance policies include a waiver by the insurers of their subrogation rights against directors and other employees of every insured except where there is fraud.

85.3 The Parties comply with the terms and conditions of the insurance policies.

85.4 Any amount not recovered from an insurer is borne by the *Employer* for events which are at his risk and by the *Contractor* for events which are at his risk.

If the *Contractor* does not insure **86**

86.1 The *Employer* may insure a risk which this contract requires the *Contractor* to insure if the *Contractor* does not submit a required certificate. The cost of this insurance to the *Employer* is paid by the *Contractor*.

Insurance by the *Employer* 87

87.1   The *Project Manager* submits policies and certificates for insurances provided by the *Employer* to the *Contractor* for acceptance before the *starting date* and afterwards as the *Contractor* instructs. The *Contractor* accepts the policies and certificates if they comply with this contract.

87.2   The *Contractor*'s acceptance of an insurance policy or certificate provided by the *Employer* does not change the responsibility of the *Employer* to provide the insurances stated in the Contract Data.

87.3   The *Contractor* may insure a risk which this contract requires the *Employer* to insure if the *Employer* does not submit a required policy or certificate. The cost of this insurance to the *Contractor* is paid by the *Employer*.

Insurance by the *Employer* 87

87.1   The *Project Manager* submits policies and certificates for insurances provided by the *Employer* to the *Contractor* for acceptance before the *starting date* and afterwards as the *Contractor* instructs. The *Contractor* accepts the policies and certificates if they comply with this contract.

87.2   The *Contractor*'s acceptance of an insurance policy or certificate provided by the *Employer* does not change the responsibility of the *Employer* to provide the insurances stated in the Contract Data.

87.3   The *Contractor* may insure a risk which this contract requires the *Employer* to insure if the *Employer* does not submit a required policy or certificate. The cost of this insurance to the *Contractor* is paid by the *Employer*.

9 Disputes and termination

Termination **94**

94.1   If either Party wishes to terminate, he notifies the *Project Manager* giving detai of his reason for terminating. The *Project Manager* issues a termination certi cate promptly if the reason complies with this contract.

94.2   The *Contractor* may terminate only for a reason identified in the Terminatic Table. The *Employer* may terminate for any reason. The procedures followe and the amounts due on termination are in accordance with the Terminatic Table.

TERMINATION TABLE

Terminating Party	Reason	Procedure	Amount due
The *Employer*	A reason other than R1 – R21	P1 and P2	A1, A2 and A4
	R1 – R15, R19	P1, P2 and P3	A1 and A3
	R17, R18, R21	P1 and P3	A1, A2 and A5
The *Contractor*	R1 - R10, R16, R20	P1 and P4	A1, A2 and A4
	R17, R18, R21	P1 and P4	A1, A2 and A5

94.3   The procedures for termination are implemented immediately after the *Proje* *Manager* has issued a termination certificate.

94.4   Within thirteen weeks of termination, the *Project Manager* certifies a final pa ment to or from the *Contractor* which is the *Project Manager*'s assessment of t amount due on termination less the total of previous payments.

94.5   After a termination certificate has been issued, the *Contractor* does no furth work necessary to complete the *works*.

Termination

ECC2 clauses on disputes have been replaced with Options W1 and W2 in ECC3.

Termination **90**

90.1   If either Party wishes to terminate the *Contractor*'s obligation to Provide the Works he notifies the *Project Manager* and the other Party giving details of his reason for terminating. The *Project Manager* issues a termination certificate to both Parties promptly if the reason complies with this contract.

Clarity is brought in the amendment that the termination is of the *Contractor*'s responsibility to Provide the Works.

90.2   The *Contractor* may terminate only for a reason identified in the Termination Table. The *Employer* may terminate for any reason. The procedures followed and the amounts due on termination are in accordance with the Termination Table.

TERMINATION TABLE

Terminating Party	Reason	Procedure	Amount due
The *Employer*	A reason other than R1–R21	P1 and P2	A1, A2 and A4
	R1–R15 or R18	P1, P2 and P3	A1 and A3
	R17 or R20	P1 and P3	A1 and A2
	R21	P1 and P4	A1 and A2
The *Contractor*	R1–R10, R16 or R19	P1 and P4	A1, A2 and A4
	R17 or R20	P1 and P4	A1 and A2

ECC2 R17 has been deleted, the existing reasons re-numbered and new R21 added. Amount due A5 has also been deleted.

90.3   The procedures for termination are implemented immediately after the *Project Manager* has issued a termination certificate.

90.4   Within thirteen weeks of termination, the *Project Manager* certifies a final payment to or from the *Contractor* which is the *Project Manager*'s assessment of the amount due on termination less the total of previous payments. Payment is made within three weeks of the *Project Manager*'s certificate.

Clarification is added that payment is made following certification of such final payment.

90.5   After a termination certificate has been issued, the *Contractor* does no further work necessary to Provide the Works.

Reasons for termination 95

95.1   Either Party may terminate if the other Party has done one of the following or its equivalent.

(1) If the other Party is an individual and has

- presented his petition for bankruptcy (R1),

- had a bankruptcy order made against him (R2),

- had a receiver appointed over his assets (R3) or

- made an arrangement with his creditors (R4).

(2) If the other Party is a company or partnership and has

- had a winding-up order made against it (R5),

- had a provisional liquidator appointed to it (R6),

- passed a resolution for winding-up (other than in order to amalgamate or reconstruct) (R9),

- had an administration order made against it (R8),

- had a receiver, receiver and manager, or administrative receiver appointed over the whole or a substantial part of its undertaking or assets (R7) or

- made an arrangement with its creditors (R10).

95.2   The *Employer* may terminate if the *Project Manager* has notified that the *Contractor* has defaulted in one of the following ways and not put the default right within four weeks of the notification.

- Substantially failed to comply with his obligations (R11).

- Not provided a bond or guarantee which this contract requires (R12).

- Appointed a Subcontractor for substantial work before the *Project Manager* has accepted the Subcontractor (R13).

95.3   The *Employer* may terminate if the *Project Manager* has notified that the *Contractor* has defaulted in one of the following ways and not stopped defaulting within four weeks of the notification.

- Substantially hindered the *Employer* or Others (R14).

- Substantially broken a health or safety regulation (R15).

95.4   The *Contractor* may terminate if the *Employer* has not paid an amount certified by the *Project Manager* within thirteen weeks of the date of the certificate (R16).

95.5   Either Party may terminate if

- war or radioactive contamination has substantially affected the *Contractor's* work for 26 weeks (R17) or

- the Parties have been released under the law from further performance of the whole of this contract (R18).

95.6   If the *Project Manager* has instructed the *Contractor* to stop or not to start any substantial work or all work and an instruction allowing the work to restart or start has not been given within thirteen weeks,

- the *Employer* may terminate if the instruction was due to a default by the *Contractor* (R19),

- the *Contractor* may terminate if the instruction was due to a default by the *Employer* (R20) and

- either Party may terminate if the instruction was due to any other reason (R21).

Reasons for termination **91**

91.1 Either Party may terminate if the other Party has done one of the following or its equivalent.

- If the other Party is an individual and has

  - presented his petition for bankruptcy (R1),
  - had a bankruptcy order made against him (R2),
  - had a receiver appointed over his assets (R3) or
  - made an arrangement with his creditors (R4).

- If the other Party is a company or partnership and has

  - had a winding-up order made against it (R5),
  - had a provisional liquidator appointed to it (R6),
  - passed a resolution for winding-up (other than in order to amalgamate or reconstruct) (R7),
  - had an administration order made against it (R8),
  - had a receiver, receiver and manager, or administrative receiver appointed over the whole or a substantial part of its undertaking or assets (R9) or
  - made an arrangement with its creditors (R10).

91.2 The *Employer* may terminate if the *Project Manager* has notified that the *Contractor* has defaulted in one of the following ways and not put the default right within four weeks of the notification.

- Substantially failed to comply with his obligations (R11).
- Not provided a bond or guarantee which this contract requires (R12).
- Appointed a Subcontractor for substantial work before the *Project Manager* has accepted the Subcontractor (R13).

91.3 The *Employer* may terminate if the *Project Manager* has notified that the *Contractor* has defaulted in one of the following ways and not stopped defaulting within four weeks of the notification.

- Substantially hindered the *Employer* or Others (R14).
- Substantially broken a health or safety regulation (R15).

91.4 The *Contractor* may terminate if the *Employer* has not paid an amount certified by the *Project Manager* within thirteen weeks of the date of the certificate (R16).

91.5 Either Party may terminate if the Parties have been released under the law from further performance of the whole of this contract (R17).

91.6 If the *Project Manager* has instructed the *Contractor* to stop or not to start any substantial work or all work and an instruction allowing the work to re-start or start has not been given within thirteen weeks,

- the *Employer* may terminate if the instruction was due to a default by the *Contractor* (R18),
- the *Contractor* may terminate if the instruction was due to a default by the *Employer* (R19) and
- either Party may terminate if the instruction was due to any other reason (R20).

Procedures on termination **96**

96.1   On termination, the *Employer* may complete the *works* himself or employ other people to do so and may use any Plant and Materials to which he has title (P1).

96.2   The procedure on termination also includes one or more of the following as set out in the Termination Table.

P2   The *Employer* may instruct the *Contractor* to leave the Site, remove any Equipment, Plant and Materials from the Site and assign the benefit of any subcontract or other contract related to performance of this contract to the *Employer*.

P3   The *Employer* may use any Equipment to which he has title.

P4   The *Contractor* leaves the Working Areas and removes the Equipment.

Payment on termination **97**

97.1   The amount due on termination includes (A1)

- an amount due assessed as for normal payments,

- the Actual Cost for Plant and Materials

    - within the Working Areas or

    - to which the *Employer* has title and of which the *Contractor* has to accept delivery,

- other Actual Cost reasonably incurred in expectation of completing the whole of the *works*,

- any amounts retained by the *Employer* and

- a deduction of any unrepaid balance of an advanced payment.

97.2   The amount due on termination also includes one or more of the following as set out in the Termination Table.

A2   The forecast Actual Cost of removing the Equipment.

A3   A deduction of the forecast of the additional cost to the *Employer* of completing the whole of the *works*.

A4   The *fee percentage* applied to

- for Options A, B, C and D, any excess of the total of the Prices at the Contract Date over the Price for Work Done to Date or

- for Options E and F, any excess of the first forecast of the Actual Cost for the *works* over the Price for Work Done to Date less the Fee.

A5   Half of A4.

91.7   The *Employer* may terminate if an event occurs which

- stops the *Contractor* completing the *works* or
- stops the *Contractor* completing the *works* by the date shown on the Accepted Programme and is forecast to delay Completion by more than 13 weeks,

and which

- neither Party could prevent and
- an experienced contractor would have judged at the Contract Date to have such a small chance of occurring that it would have been unreasonable for him to have allowed for it (R21).

> The additional clause 19 'prevention' provisions are also now a reason for termination by the *Employer*. This change replaces a reason for termination being war or radioactive contamination (ECC2 R17).

Procedures on termination   **92**

92.1   On termination, the *Employer* may complete the *works* and may use any Plant and Materials to which he has title (P1).

92.2   The procedure on termination also includes one or more of the following as set out in the Termination Table.

P2   The *Employer* may instruct the *Contractor* to leave the Site, remove any Equipment, Plant and Materials from the Site and assign the benefit of any subcontract or other contract related to performance of this contract to the *Employer*.

P3   The *Employer* may use any Equipment to which the *Contractor* has title to complete the *works*. The *Contractor* promptly removes the Equipment from Site when the *Project Manager* notifies him that the *Employer* no longer requires it to complete the *works*.

P4   The *Contractor* leaves the Working Areas and removes the Equipment.

Payment on termination   **93**

93.1   The amount due on termination includes (A1)

- an amount due assessed as for normal payments,
- the Defined Cost for Plant and Materials
  - within the Working Areas or
  - to which the *Employer* has title and of which the *Contractor* has to accept delivery,
- other Defined Cost reasonably incurred in expectation of completing the whole of the *works*,
- any amounts retained by the *Employer* and
- a deduction of any un-repaid balance of an advanced payment.

93.2   The amount due on termination also includes one or more of the following as set out in the Termination Table.

A2   The forecast Defined Cost of removing the Equipment.

A3   A deduction of the forecast of the additional cost to the *Employer* of completing the whole of the *works*.

A4   The *direct fee percentage* applied to

- for Options A, B, C and D, any excess of the total of the Prices at the Contract Date over the Price for Work Done to Date or
- for Options E and F, any excess of the first forecast of the Defined Cost for the *works* over the Price for Work Done to Date less the Fee.

MAIN OPTION CLAUSES

Option A: Priced contract with activity schedule

Identified and defined terms 11

11.2   (28) Actual Cost is the cost of the components in the Schedule of Cost Components whether work is subcontracted or not excluding the cost of preparing quotations for compensation events.

(24) The Price for Work Done to Date is the total of the Prices for

- each group of completed activities and
- each completed activity which is not in a group

which is without Defects which would either delay or be covered by immediately following work.

(20) The Prices are the lump sum prices for each of the activities in the *activity schedule* unless later changed in accordance with this contract.

The programme 31

31.4   The *Contractor* shows the start and finish of each activity on the *activity schedule* on each programme which he submits for acceptance.

Acceleration 36

36.3   When the *Project Manager* accepts a quotation for an acceleration, he changes the Completion Date and the Prices accordingly and accepts the revised programme.

MAIN OPTION CLAUSES

Option A: Priced contract with activity schedule

Where similar changes are made within the different main Options, the commentary is not repeated.

Identified and defined terms

11

11.2   (20) The Activity Schedule is the *activity schedule* unless later changed in accordance with this contract.

'Activity Schedule' is now a defined term and uses the '*activity schedule*' included in the Contract Data part two as this document unless later changed, for example by the implementation of a compensation event.

(22) Defined Cost is the cost of the components in the Shorter Schedule of Cost Components whether work is subcontracted or not excluding the cost of preparing quotations for compensation events.

Apart from the change of 'Actual Cost' to 'Defined Cost', the definition of Defined Cost now has regard to the cost of the components in the Shorter Schedule of Cost Components rather than the Schedule of Cost Components as ECC2. It is important for users to appreciate the differences between the two Schedules and their relationships with each of the main Options.

(27) The Price for Work Done to Date is the total of the Prices for

- each group of completed activities and
- each completed activity which is not in a group.

A completed activity is one which is without Defects which would either delay or be covered by immediately following work.

A minor amendment to clarify that the final sentence should be read in relation to completed activities.

(30) The Prices are the lump sum prices for each of the activities on the Activity Schedule unless later changed in accordance with this contract.

The programme

31

31.4   The *Contractor* provides information which shows how each activity on the Activity Schedule relates to the operations on each programme which he submits for acceptance.

Instead of the ECC2 approach of showing the start and finish of each activity on each programme, the *Contractor* provides information showing how each activity relates to the operations on each programme.

Acceleration

36

36.3   When the *Project Manager* accepts a quotation for an acceleration, he changes the Prices, the Completion Date and the Key Dates accordingly and accepts the revised programme.

Any Key Dates can now be changed where they are part of an acceleration agreement.

           69

The *activity schedule* **54**

54.1 Information in the *activity schedule* is not Works Information or Site Information.

54.2 If the *Contractor* changes a planned method of working at his discretion so that the *activity schedule* does not comply with the Accepted Programme, he submits a revision of the *activity schedule* to the *Project Manager* for acceptance.

54.3 A reason for not accepting a revision of the *activity schedule* is that

- it does not comply with the Accepted Programme,
- any changed Prices are not reasonably distributed between the activities or
- the total of the Prices is changed.

Assessing compensation **63**
events 63.2 If the effect of a compensation event is to reduce the total Actual Cost and the event is

- a change to the Works Information or
- a correction of an assumption stated by the *Project Manager* for assessing an earlier compensation event,

the Prices are reduced.

63.8 Assessments for changed Prices for compensation events are in the form of changes to the *activity schedule*.

63.10 The assessment of a compensation event which is or includes subcontracted work has the *Contractor*'s *fee percentage* added to Actual Cost but fees paid or to be paid by the *Contractor* to the Subcontractors are not added.

63.11 If the *Project Manager* and the *Contractor* agree, the *Contractor* assesses a compensation event using the Shorter Schedule of Cost Components. The *Project Manager* may make his own assessments using the Shorter Schedule of Cost Components.

Implementing compensation **65**
events 65.4 The *Project Manager* includes the changes to the Prices and the Completion Date from the quotation which he has accepted or from his own assessment in his notification implementing a compensation event.

Payment on termination **97**
97.3 The amount due on termination is assessed without taking grouping of activities into account.

The Activity Schedule **54**

54.1 Information in the Activity Schedule is not Works Information or Site Information.

54.2 If the *Contractor* changes a planned method of working at his discretion so that the activities on the Activity Schedule do not relate to the operations on the Accepted Programme, he submits a revision of the Activity Schedule to the *Project Manager* for acceptance.

> Rather than the ECC2 approach of all activities on the *activity schedule* being shown on each programme, ECC3 reinforces the programme as an operational tool with information being provided that links the Activity Schedule activities to the operations on each programme.

54.3 A reason for not accepting a revision of the Activity Schedule is that

- it does not comply with the Accepted Programme,
- any changed Prices are not reasonably distributed between the activities or
- the total of the Prices is changed.

Assessing compensation 63
events

63.10 If the effect of a compensation event is to reduce the total Defined Cost and the event is

- a change to the Works Information or
- a correction of an assumption stated by the *Project Manager* for assessing an earlier compensation event,

the Prices are reduced.

> This subclause has been relocated from core clause 63.2 and modified slightly to suit Options A and B.

> ECC2 subclauses 63.10 and 63.11 for both Options A and B are deleted. Subclause 63.10 is covered in the Shorter Schedule of Cost Components' opening paragraph and the option to use the Schedule of Cost Components has now been deleted in favour of only using the simplified Shorter Schedule.

63.12 Assessments for changed Prices for compensation events are in the form of changes to the Activity Schedule.

63.14 If the *Project Manager* and the *Contractor* agree, rates and lump sums may be used to assess a compensation event instead of Defined Cost.

> By agreement, this new subclause can be used to progress the assessment of compensation events more rapidly by using rates and lump sums instead of Defined Cost. Where applicable, this ought to speed the process up and therefore reduce time and costs in dealing with compensation events.

Implementing 65
compensation events

65.4 The changes to the Prices, the Completion Date and the Key Dates are included in the notification implementing a compensation event.

Payment on termination 93

93.3 The amount due on termination is assessed without taking grouping of activities into account.

Option B: Priced contract with bill of quantities

Identified and defined terms **11**

11.2 (28) Actual Cost is the cost of the components in the Schedule of Cost Components whether work is subcontracted or not excluding the cost of preparing quotations for compensation events.

(25) The Price for Work Done to Date is the total of

- the quantity of the work which the *Contractor* has completed for each item in the *bill of quantities* multiplied by the rate and
- a proportion of each lump sum which is the proportion of the work covered by the item which the *Contractor* has completed.

In this clause, completed work means work without Defects which would either delay or be covered by immediately following work.

(21) The Prices are the lump sums and the amounts obtained by multiplying the rates by the quantities for the items in the *bill of quantities* unless later changed in accordance with this contract.

Acceleration **36**

36.3 When the *Project Manager* accepts a quotation for an acceleration, he changes the Completion Date and the Prices accordingly and accepts the revised programme.

The *bill of quantities* **55**

55.1 Information in the *bill of quantities* is not Works Information or Site Information.

Compensation events **60**

60.4 A difference between the final total quantity of work done and the quantity stated for an item in the *bill of quantities* at the Contract Date is a compensation event if

- the difference causes the Actual Cost per unit of quantity to change and
- the rate in the *bill of quantities* for the item at the Contract Date multiplied by the final total quantity of work done is more than 0.1% of the total of the Prices at the Contract Date.

If the Actual Cost per unit of quantity is reduced, the affected rate is reduced.

60.5 A difference between the final total quantity of work done and the quantity for an item stated in the *bill of quantities* at the Contract Date which delays Completion is a compensation event.

Option B: Priced contract with bill of quantities

Identified and defined terms **11**

11.2   (21) The Bill of Quantities is the *bill of quantities* as changed in accordance with this contract to accommodate implemented compensation events and for accepted quotations for acceleration.

> 'Bill of Quantities' has now been introduced as a defined term in a similar manner to how the contract deals with the 'Activity Schedule' under main Options A and C.

(22) Defined Cost is the cost of the components in the Shorter Schedule of Cost Components whether work is subcontracted or not excluding the cost of preparing quotations for compensation events.

> As with main Option A, the Shorter Schedule of Cost Components is used to assess Defined Cost.

(28) The Price for Work Done to Date is the total of

- the quantity of the work which the *Contractor* has completed for each item in the Bill of Quantities multiplied by the rate and
- a proportion of each lump sum which is the proportion of the work covered by the item which the *Contractor* has completed.

Completed work is work without Defects which would either delay or be covered by immediately following work.

(31) The Prices are the lump sums and the amounts obtained by multiplying the rates by the quantities for the items in the Bill of Quantities.

Acceleration **36**

36.3   When the *Project Manager* accepts a quotation for an acceleration, he changes the Prices, the Completion Date and the Key Dates accordingly and accepts the revised programme.

The Bill of Quantities **55**

55.1   Information in the Bill of Quantities is not Works Information or Site Information.

Compensation events **60**

60.4   A difference between the final total quantity of work done and the quantity stated for an item in the Bill of Quantities is a compensation event if

- the difference does not result from a change to the Works Information,
- the difference causes the Defined Cost per unit of quantity to change and
- the rate in the Bill of Quantities for the item multiplied by the final total quantity of work done is more than 0.5% of the total of the Prices at the Contract Date.

If the Defined Cost per unit of quantity is reduced, the affected rate is reduced.

> An additional test in this subclause is that the difference in quantities must not result from a change to the Works Information. The last test of whether such a matter consti- tutes a compensation event has also changed. For this to be a compensation event in ECC2 one took the rate in the *bill of quantities* for the item at the Contract Date multiplied by the final quantity of work done and the product was required to be more than 0.1% of the total of the Prices in ECC2. This has now been raised to 0.5% for the rate in the Bill of Quantities in ECC3.

60.5   A difference between the final total quantity of work done and the quantity for an item stated in the Bill of Quantities which delays Completion or the meeting of the Condition stated for a Key Date is a compensation event.

> An increase in quantities that affects the meeting of a Condition stated for a Key Date can also now constitute grounds for a compensation event.

60.6   The *Project Manager* corrects mistakes in the bill of quantities which are depa tures from the *method of measurement* or are due to ambiguities or inconsiste cies. Each such correction is a compensation event which may lead to reduce Prices.

Assessing compensation **63**
events   63.2   If the effect of a compensation event is to reduce the total Actual Cost and th event is

- a change to the Works Information or
- a correction of an assumption stated by the *Project Manager* for assessir an earlier compensation event,

the Prices are reduced.

63.9   Assessments for changed Prices for compensation events are in the form changes to the *bill of quantities*. If the *Project Manager* and the *Contract* agree, rates and lump sums in the *bill of quantities* may be used as a basis f assessment instead of Actual Cost and the resulting Fee.

63.10   The assessment of a compensation event which is or includes subcontracte work has the *Contractor*'s *fee percentage* added to Actual Cost but fees paid to be paid by the *Contractor* to the Subcontractors are not added.

63.11   If the *Project Manager* and the *Contractor* agree, the *Contractor* assesses a con pensation event using the Shorter Schedule of Cost Components. The *Proje Manager* may make his own assessments using the Shorter Schedule of Co Components.

Implementing compensation **65**
events   65.4   The *Project Manager* includes the changes to the Prices and the Completic Date from the quotation which he has accepted or from his own assessment his notification implementing a compensation event.

60.6   The *Project Manager* corrects mistakes in the Bill of Quantities which are departures from the rules for item descriptions and for division of the work into items in the *method of measurement* or are due to ambiguities or inconsistencies. Each such correction is a compensation event which may lead to reduced Prices.

Further clarification has been added in this subclause. It is departures from the rules for item descriptions and for division of the work into items in the *method of measurement* that constitutes grounds for a compensation event.

60.7   In assessing a compensation event which results from a correction of an inconsistency between the Bill of Quantities and another document, the *Contractor* is assumed to have taken the Bill of Quantities as correct.

This new subclause confirms the basis of pricing where a compensation event arises from a correction of an inconsistency between the Bill of Quantities and another document.

Assessing compensation **63**
events 63.10 If the effect of a compensation event is to reduce the total Defined Cost and the event is

- a change to the Works Information or
- a correction of an assumption stated by the *Project Manager* for assessing an earlier compensation event,

the Prices are reduced.

63.13 Assessments for changed Prices for compensation events are in the form of changes to the Bill of Quantities.

- For the whole or a part of a compensation event for work not yet done and for which there is an item in the Bill of Quantities, the changes are

  - a changed rate,
  - a changed quantity or
  - a changed lump sum.

- For the whole or a part of a compensation event for work not yet done and for which there is no item in the Bill of Quantities, the change is a new priced item which, unless the *Project Manager* and the *Contractor* agree otherwise, is compiled in accordance with the *method of measurement*.
- For the whole or a part of a compensation event for work already done, the change is a new lump sum item.

If the *Project Manager* and the *Contractor* agree, rates and lump sums may be used to assess a compensation event instead of Defined Cost.

The amendments to this subclause detail the form of changes that are made to the Bill of Quantities resulting from compensation events. The form depends on whether the work is not yet done or already done, and if not yet done whether there is or isn't an item in the Bill of Quantities. As with Option A, there is now provided an alternative for assessing compensation events, by agreement, by using rates and lump sums instead of Defined Cost.

Implementing **65**
compensation events 65.4 The changes to the Prices, the Completion Date and the Key Dates are included in the notification implementing a compensation event.

Option C: Target contract with activity schedule

Identified and defined terms **11**

11.2 (27) Actual Cost is the amount of payments due to Subcontractors for wor which is subcontracted and the cost of the components in the Schedule of Co Components for work which is not subcontracted, less any Disallowed Cost.

(30) Disallowed Cost is cost which the *Project Manager* decides

- is not justified by the *Contractor*'s accounts and records,
- should not have been paid to a Subcontractor in accordance with h subcontract,
- was incurred only because the *Contractor* did not
  - follow an acceptance or procurement procedure stated in the Worl Information or
  - give an early warning which he could have given or
- results from paying a Subcontractor more for a compensation event than included in the accepted quotation or assessment for the compensatic event

and the cost of

- correcting Defects after Completion,
- correcting Defects caused by the *Contractor* not complying with a requir ment for how he is to Provide the Works stated in the Works Informatior
- Plant and Materials not used to Provide the Works (after allowing f reasonable wastage) and
- resources not used to Provide the Works (after allowing for reasonab availability and utilisation) or not taken away from the Working Are when the *Project Manager* requested.

Option C: Target contract with activity schedule

Identified and defined   **11**
terms   11.2   (20) The Activity Schedule is the *activity schedule* unless later changed in accordance with this contract.

(23) Defined Cost is

- the amount of payments due to Subcontractors for work which is subcontracted without taking account of amounts deducted for

  - retention,
  - payment to the *Employer* as a result of the Subcontractor failing to meet a Key Date,
  - the correction of Defects after Completion,
  - payments to Others and
  - the supply of equipment, supplies and services included in the charge for overhead cost within the Working Areas in this contract

and

- the cost of components in the Schedule of Cost Components for other work

less Disallowed Cost.

> When assessing Defined Cost in terms of Subcontractors, there is now a list of matters that are to be ignored in this process. This is designed to prevent the *Contractor* from in effect suffering two lots of deductions for example for retention or correcting Defects after Completion. In ECC2, if retention was held by both *Employer* and *Contractor*, the *Employer* paid the *Contractor* the net amount the *Contractor* paid the Subcontractor and then took a further retention deduction from the *Contractor*. Whilst this was rectified when retention monies were finally released, this would not be the case for instances such as correcting Defects after Completion.

(25) Disallowed Cost is cost which the *Project Manager* decides

- is not justified by the *Contractor*'s accounts and records,
- should not have been paid to a Subcontractor or supplier in accordance with his contract,
- was incurred only because the *Contractor* did not

  - follow an acceptance or procurement procedure stated in the Works Information or
  - give an early warning which this contract required him to give

and the cost of

- correcting Defects after Completion,
- correcting Defects caused by the *Contractor* not complying with a constraint on how he is to Provide the Works stated in the Works Information,
- Plant and Materials not used to Provide the Works (after allowing for reasonable wastage) unless resulting from a change to the Works Information,

(23) The Price for Work Done to Date is the Actual Cost which the *Contractor* has paid plus the Fee.

(20) The Prices are the lump sum prices for each of the activities in the *activity schedule* unless later changed in accordance with this contract.

Providing the Works 20

20.3   The *Contractor* advises the *Project Manager* on the practical implications of the design of the *works* and on subcontracting arrangements.

20.4   The *Contractor* prepares forecasts of the total Actual Cost for the whole of the *works* in consultation with the *Project Manager* and submits them to the *Project Manager*. Forecasts are prepared at the intervals stated in the Contract Data from the *starting date* until Completion of the whole of the *works*. An explanation of the changes made since the previous forecast is submitted with each forecast.

Subcontracting 26

26.4   The *Contractor* submits the proposed contract data for each subcontract for acceptance to the *Project Manager* if

- the NEC Engineering and Construction Subcontract or the NEC Professional Services Contract is to be used and
- the *Project Manager* instructs the *Contractor* to make the submission.

A reason for not accepting the proposed contract data is that its use will not allow the *Contractor* to Provide the Works.

The programme 31

31.4   The *Contractor* shows the start and finish of each activity on the *activity schedule* on each programme which he submits for acceptance.

- resources not used to Provide the Works (after allowing for reasonable availability and utilisation) or not taken away from the Working Areas when the *Project Manager* requested and
- preparation for and conduct of an adjudication or proceedings of the *tribunal*.

There are a number of amendments, additions and clarifications to this subclause. There is clarification that costs that should not have been paid to 'a supplier' in accordance with his contract can be a Disallowed Cost. In relation to Disallowed Costs, these must arise from early warnings the *Contractor* is required to give under this contract. In terms of correcting Defects, the failure to comply with a 'requirement' for how the *Contractor* is to Provide the Works is altered to 'constraint'. This is consistent with the defined term of Works Information. Finally, preparation for and conduct of an adjudication or proceedings of the *tribunal* are clearly a Disallowed Cost.

(29) The Price for Work Done to Date is the total Defined Cost which the *Project Manager* forecasts will have been paid by the *Contractor* before the next assessment date plus the Fee.

A major change to this subclause sees the *Contractor* recovering Defined Costs for each assessment based on what the *Project Manager* forecasts will have been paid by the *Contractor* before the next assessment date. This moves this Option towards a cash neutral basis rather than the more likely cash negative basis that exists under ECC2, where only the Actual Cost paid by the *Contractor* was used in the assessment.

(30) The Prices are the lump sum prices for each of the activities on the Activity Schedule unless later changed in accordance with this contract.

Providing the Works 20

20.3   The *Contractor* advises the *Project Manager* on the practical implications of the design of the *works* and on subcontracting arrangements.

20.4   The *Contractor* prepares forecasts of the total Defined Cost for the whole of the *works* in consultation with the *Project Manager* and submits them to the *Project Manager*. Forecasts are prepared at the intervals stated in the Contract Data from the *starting date* until Completion of the whole of the *works*. An explanation of the changes made since the previous forecast is submitted with each forecast.

Subcontracting 26

26.4   The *Contractor* submits the proposed contract data for each subcontract for acceptance to the *Project Manager* if

- an NEC contract is proposed and
- the *Project Manager* instructs the *Contractor* to make the submission.

A reason for not accepting the proposed contract data is that its use will not allow the *Contractor* to Provide the Works.

Recognising the increasing use of NEC contracts in subcontract appointments, the first bullet has been amended to 'NEC contracts' rather than just the two named NEC contracts under these provisions in ECC2.

The programme 31

31.4   The *Contractor* provides information which shows how each activity on the Activity Schedule relates to the operations on each programme which he submits for acceptance.

Acceleration **36**

36.3 When the *Project Manager* accepts a quotation for an acceleration, he change the Completion Date and the Prices accordingly and accepts the revised pro gramme.

36.5 The *Contractor* submits a Subcontractor's proposal to accelerate to the *Projec Manager* for acceptance.

Assessing the amount due **50**

50.6 Payments of Actual Cost made by the *Contractor* in a currency other than th *currency of this contract* are included in the amount due as payments to be mad to him in the same currency. Such payments are converted to the *currency c this contract* in order to calculate the Fee and any *Contractor*'s share using th *exchange rates*.

Actual Cost **52**

52.2 The *Contractor* keeps

- accounts of his payments of Actual Cost,
- records which show that the payments have been made,
- records of communications and calculations relating to assessment o compensation events for Subcontractors and
- other accounts and records as stated in the Works Information.

52.3 The *Contractor* allows the *Project Manager* to inspect at any time withir working hours the accounts and records which he is required to keep.

The *Contractor*'s share **53**

53.1 The *Project Manager* assesses the *Contractor*'s share of the difference betweer the total of the Prices and the Price for Work Done to Date. The difference i divided into increments falling within each of the *share ranges*. The limits of a *share range* are a Price for Work Done to Date divided by the total of the Prices expressed as a percentage. The *Contractor*'s share equals the sum of the product of the increment within each *share range* and the corresponding *Contractor*' *share percentage*.

53.2 If the Price for Work Done to Date is less than the total of the Prices, the *Contractor* is paid his share of the saving. If the Price for Work Done to Date i greater than the total of the Prices, the *Contractor* pays his share of the excess.

53.3 The *Project Manager* assesses the *Contractor*'s share at Completion of the whole of the *works* using his forecasts of the final Price for Work Done to Date and the final total of the Prices. This share is included in the amount due following Completion of the whole of the *works*.

Acceleration   **36**

36.3   When the *Project Manager* accepts a quotation for an acceleration, he changes the Prices, the Completion Date and the Key Dates accordingly and accepts the revised programme.

Tests and inspections   **40**

40.7   When the *Project Manager* assesses the cost incurred by the *Employer* in repeating a test or inspection after a Defect is found, the *Project Manager* does not include the *Contractor*'s cost of carrying out the repeat test or inspection.

> This new subclause clarifies that that additional costs the *Employer* incurs arising from a repeat test or inspection due to a Defect do not include the *Contractor*'s costs. In most cases before Completion, the *Contractor* recovers such costs as a Defined Cost, which ultimately reduces the amount of potential Contractor's share.

ssessing the amount due   **50**

50.6   Payments of Defined Cost made by the *Contractor* in a currency other than the *currency of this contract* are included in the amount due as payments to be made to him in the same currency. Such payments are converted to the *currency of this contract* in order to calculate the Fee and any *Contractor*'s share using the *exchange rates*.

Defined Cost   **52**

52.2   The *Contractor* keeps these records

- accounts of payments of Defined Cost,
- proof that the payments have been made,
- communications about and assessments of compensation events for Subcontractors and
- other records as stated in the Works Information.

> Minor clarifying amendments to this subclause includes all four bulleted items being considered as 'records' and now the *Contractor* is to keep 'proof' that payments have been made.

52.3   The *Contractor* allows the *Project Manager* to inspect at any time within working hours the accounts and records which he is required to keep.

The *Contractor*'s share   **53**

53.1   The *Project Manager* assesses the *Contractor*'s share of the difference between the total of the Prices and the Price for Work Done to Date. The difference is divided into increments falling within each of the *share ranges*. The limits of a *share range* are the Price for Work Done to Date divided by the total of the Prices, expressed as a percentage. The *Contractor*'s share equals the sum of the products of the increment within each *share range* and the corresponding *Contractor*'s *share percentage*.

53.2   If the Price for Work Done to Date is less than the total of the Prices, the *Contractor* is paid his share of the saving. If the Price for Work Done to Date is greater than the total of the Prices, the *Contractor* pays his share of the excess.

53.3   The *Project Manager* makes a preliminary assessment of the *Contractor*'s share at Completion of the whole of the *works* using his forecasts of the final Price for Work Done to Date and the final total of the Prices. This share is included in the amount due following Completion of the whole of the *works*.

53.4   The *Project Manager* again assesses the *Contractor*'s share using the final Pri
for Work Done to Date and the final total of the Prices. This share is include
in the final amount due.

53.5   If the *Project Manager* accepts a proposal by the *Contractor* to change tl
Works Information provided by the *Employer* so that the Actual Cost
reduced, the Prices are not reduced.

The *activity schedule* 54

54.1   Information in the *activity schedule* is not Works Information or Site Inform
tion.

54.2   If the *Contractor* changes a planned method of working at his discretion so th
the *activity schedule* does not comply with the Accepted Programme, he submi
a revision of the *activity schedule* to the *Project Manager* for acceptance.

54.3   A reason for not accepting a revision of the *activity schedule* is that

- it does not comply with the Accepted Programme,

- any changed Prices are not reasonably distributed between the activities o

- the total of the Prices is changed.

**Assessing compensation 63
events** 63.2   If the effect of a compensation event is to reduce the total Actual Cost and tl
event is

- a change to the Works Information or

- a correction of an assumption stated by the *Project Manager* for assessir
an earlier compensation event,

the Prices are reduced.

63.8   Assessments for changed Prices for compensation events are in the form o
changes to the *activity schedule*.

63.11   If the *Project Manager* and the *Contractor* agree, the *Contractor* assesses a com
pensation event using the Shorter Schedule of Cost Components. The *Proje*
Manager may make his own assessments using the Shorter Schedule of Co
Components.

**Implementing compensation 65
events** 65.4   The *Project Manager* includes the changes to the Prices and the Completio
Date from the quotation which he has accepted or from his own assessment i
his notification implementing a compensation event.

53.4   The *Project Manager* makes a final assessment of the *Contractor*'s share using the final Price for Work Done to Date and the final total of the Prices. This share is included in the final amount due.

A distinction is made between subclauses 53.4 and 53.5 that subclause 53.4 is a preliminary assessment and is made using forecasts whereas the assessment under subclause 53.5 is final and uses the final Price for Word Done to Date and the final total of the Prices. ECC2 subclause 53.5 is relocated to subclause 63.11 in ECC3.

The Activity Schedule   **54**

54.1   Information in the Activity Schedule is not Works Information or Site Information.

54.2   If the *Contractor* changes a planned method of working at his discretion so that the activities on the Activity Schedule do not relate to the operations on the Accepted Programme, he submits a revision of the Activity Schedule to the *Project Manager* for acceptance.

54.3   A reason for not accepting a revision of the Activity Schedule is that

- it does not comply with the Accepted Programme,
- any changed Prices are not reasonably distributed between the activities or
- the total of the Prices is changed.

Assessing compensation   63
events   63.11   If the effect of a compensation event is to reduce the total Defined Cost and the event is

- a change to the Works Information, other than a change to the Works Information provided by the *Employer* which the *Contractor* proposed and the *Project Manager* has accepted or
- a correction of an assumption stated by the *Project Manager* for assessing an earlier compensation event,

the Prices are reduced.

This subclause has been relocated from core subclause 63.2 and modified to suit Options C, D and E. The additional text in the first bullet provides for what is often called *Contractor*'s value engineering proposals by leaving the Prices as they are and the Parties deriving financial benefit through the share provisions.

63.12   Assessments for changed Prices for compensation events are in the form of changes to the Activity Schedule.

63.15   If the *Project Manager* and the *Contractor* agree, the *Contractor* assesses a compensation event using the Shorter Schedule of Cost Components. The *Project Manager* may make his own assessments using the Shorter Schedule of Cost Components.

Implementing   65
compensation events   65.4   The changes to the Prices, the Completion Date and the Key Dates are included in the notification implementing a compensation event.

Payment on termination 97

97.4   If there is a termination, the *Project Manager* assesses the *Contractor*'s sha
after he has certified termination. His assessment uses the Price for Work Do
to Date at termination and the total of the Prices for the work done befo
termination.

Payment on termination   **93**

93.4   If there is a termination, the *Project Manager* assesses the *Contractor*'s share after he has certified termination. His assessment uses, as the Price for Work Done to Date, the total of the Defined Cost which the *Contractor* has paid and which he is committed to pay for **work done before termination.** The assessment uses as the total of the Prices

- the lump sum price for each activity which has been completed and
- a proportion of the lump sum price for each incomplete activity which is the proportion of the work in the activity which has been completed.

An extended set of provisions for the event of termination includes now those monies the *Contractor* is committed to pay for work done before termination. It also details the basis of calculating the total of the Prices for the purposes of calculating the *Contractor*'s share.

93.6   The *Project Manager*'s assessment of the *Contractor*'s share is added to the amount due to the *Contractor* on termination if there has been a saving or deducted if there has been an excess.

Option D: Target contract with bill of quantities

Identified and defined terms   11

11.2   (27) Actual Cost is the amount of payments due to Subcontractors for work which is subcontracted and the cost of the components in the Schedule of Cost Components for work which is not subcontracted, less any Disallowed Cost.

(30) Disallowed Cost is cost which the *Project Manager* decides

- is not justified by the *Contractor*'s accounts and records,

- should not have been paid to a Subcontractor in accordance with his subcontract,

- was incurred only because the *Contractor* did not

    - follow an acceptance or procurement procedure stated in the Works Information or

    - give an early warning which he could have given or

- results from paying a Subcontractor more for a compensation event than included in the accepted quotation or assessment for the compensation event

and the cost of

- correcting Defects after Completion,

- correcting Defects caused by the *Contractor* not complying with a requirement for how he is to Provide the Works stated in the Works Information,

- Plant and Materials not used to Provide the Works (after allowing for reasonable wastage) and

- resources not used to Provide the Works (after allowing for reasonable availability and utilisation) or not taken away from the Working Areas when the *Project Manager* requested.

(23) The Price for Work Done to Date is the Actual Cost which the *Contractor* has paid plus the Fee.

(21) The Prices are the lump sums and the amounts obtained by multiplying the rates by the quantities for the items in the *bill of quantities* unless later changed in accordance with this contract.

Option D: Target contract with bill of quantities

Identified and defined **11**
terms **11.2**

(21) The Bill of Quantities is the *bill of quantities* as changed in accordance with this contract to accommodate implemented compensation events and for accepted quotations for acceleration.

(23) Defined Cost is

- the amount of payments due to Subcontractors for work which is subcontracted without taking account of amounts deducted for

  - retention,
  - payment to the *Employer* as a result of the Subcontractor failing to meet a Key Date,
  - the correction of Defects after Completion,
  - payments to Others and
  - the supply of equipment, supplies and services included in the charge for overhead cost within the Working Areas in this contract

and

- the cost of components in the Schedule of Cost Components for other work

less Disallowed Cost.

(25) Disallowed Cost is cost which the *Project Manager* decides

- is not justified by the *Contractor*'s accounts and records,
- should not have been paid to a Subcontractor or supplier in accordance with his contract,
- was incurred only because the *Contractor* did not

  - follow an acceptance or procurement procedure stated in the Works Information or
  - give an early warning which this contract required him to give

and the cost of

- correcting Defects after Completion,
- correcting Defects caused by the *Contractor* not complying with a constraint on how he is to Provide the Works stated in the Works Information,
- Plant and Materials not used to Provide the Works (after allowing for reasonable wastage) unless resulting from a change to the Works Information,
- resources not used to Provide the Works (after allowing for reasonable availability and utilisation) or not taken away from the Working Areas when the *Project Manager* requested and
- preparation for and conduct of an adjudication or proceedings of the *tribunal*.

(29) The Price for Work Done to Date is the total Defined Cost which the *Project Manager* forecasts will have been paid by the *Contractor* before the next assessment date plus the Fee.

(31) The Prices are the lump sums and the amounts obtained by multiplying the rates by the quantities for the items in the Bill of Quantities.

The definition of the Prices includes reference to the Bill of Quantities, which in turn is the tendered *bill of quantities* unless later changed, for example due to compensation events.

Providing the Works 20

20.3   The *Contractor* advises the *Project Manager* on the practical implications of the design of the *works* and on subcontracting arrangements.

20.4   The *Contractor* prepares forecasts of the total Actual Cost for the whole of the *works* in consultation with the *Project Manager* and submits them to the *Project Manager*. Forecasts are prepared at the intervals stated in the Contract Data from the *starting date* until Completion of the whole of the *works*. An explanation of the changes made since the previous forecast is submitted with each forecast.

Subcontracting 26

26.4   The *Contractor* submits the proposed contract data for each subcontract for acceptance to the *Project Manager* if

- the NEC Engineering and Construction Subcontract or the NEC Professional Services Contract is to be used and

- the *Project Manager* instructs the *Contractor* to make the submission.

A reason for not accepting the proposed contract data is that its use will not allow the *Contractor* to Provide the Works.

Acceleration 36

36.3   When the *Project Manager* accepts a quotation for an acceleration, he changes the Completion Date and the Prices accordingly and accepts the revised programme.

36.5   The *Contractor* submits a Subcontractor's proposal to accelerate to the *Project Manager* for acceptance.

Assessing the amount due 50

50.6   Payments of Actual Cost made by the *Contractor* in a currency other than the *currency of this contract* are included in the amount due as payments to be made to him in the same currency. Such payments are converted to the *currency of this contract* in order to calculate the Fee and any *Contractor*'s share using the *exchange rates*.

(33) The Total of the Prices is the total of

- the quantity of the work which the *Contractor* has completed for each item in the Bill of Quantities multiplied by the rate and
- a proportion of each lump sum which is the proportion of the work covered by the item which the *Contractor* has completed.

Completed work is work without Defects which would either delay or be covered by immediately following work.

'Total of the Prices' is a new defined term that appears in ECC3. This is quite different to 'total of the Prices' in ECC2, which was calculated using the tendered *bill of quantities* and any compensation events. This resulted in any difference in quantities not arising through compensation events not being considered within the 'total of the Prices'. In ECC3 such differences in quantities are part of the 'Total of the Prices', however this 'total' is now not known until the final quantity of the work that the *Contractor* has completed is re-measured. The *Employer* is therefore taking the risk of such errors in quantities in ECC3 whereas it was the *Contractor* who took this risk in ECC2. This term is referred to and used in calculating the *Contractor*'s share under subclauses 53.5 to 53.8.

Providing the Works   **20**

20.3   The *Contractor* advises the *Project Manager* on the practical implications of the design of the *works* and on subcontracting arrangements.

20.4   The *Contractor* prepares forecasts of the total Defined Cost for the whole of the *works* in consultation with the *Project Manager* and submits them to the *Project Manager*. Forecasts are prepared at the intervals stated in the Contract Data from the *starting date* until Completion of the whole of the *works*. An explanation of the changes made since the previous forecast is submitted with each forecast.

Subcontracting   **26**

26.4   The *Contractor* submits the proposed contract data for each subcontract for acceptance to the *Project Manager* if

- an NEC contract is proposed and
- the *Project Manager* instructs the *Contractor* to make the submission.

A reason for not accepting the proposed contract data is that its use will not allow the *Contractor* to Provide the Works.

Acceleration   **36**

36.3   When the *Project Manager* accepts a quotation for an acceleration, he changes the Prices, the Completion Date and the Key Dates accordingly and accepts the revised programme.

Tests and inspections   **40**

40.7   When the *Project Manager* assesses the cost incurred by the *Employer* in repeating a test or inspection after a Defect is found, the *Project Manager* does not include the *Contractor*'s cost of carrying out the repeat test or inspection.

ssessing the amount due   **50**

50.6   Payments of Defined Cost made by the *Contractor* in a currency other than the *currency of this contract* are included in the amount due as payments to be made to him in the same currency. Such payments are converted to the *currency of this contract* in order to calculate the Fee and any *Contractor*'s share using the *exchange rates*.

Actual Cost **52**

52.2 The *Contractor* keeps

- accounts of his payments of Actual Cost,
- records which show that the payments have been made,
- records of communications and calculations relating to assessment of compensation events for Subcontractors and
- other accounts and records as stated in the Works Information.

52.3 The *Contractor* allows the *Project Manager* to inspect at any time within working hours the accounts and records which he is required to keep.

The *Contractor*'s share **53**

53.1 The *Project Manager* assesses the *Contractor*'s share of the difference between the total of the Prices and the Price for Work Done to Date. The difference is divided into increments falling within each of the *share ranges*. The limits of a *share range* are a Price for Work Done to Date divided by the total of the Prices, expressed as a percentage. The *Contractor*'s share equals the sum of the products of the increment within each *share range* and the corresponding *Contractor*'s *share percentage*.

53.2 If the Price for Work Done to Date is less than the total of the Prices, the *Contractor* is paid his share of the saving. If the Price for Work Done to Date is greater than the total of the Prices, the *Contractor* pays his share of the excess.

53.3 The *Project Manager* assesses the *Contractor*'s share at Completion of the whole of the *works* using his forecasts of the final Price for Work Done to Date and the final total of the Prices. This share is included in the amount due following Completion of the whole of the *works*.

53.4 The *Project Manager* again assesses the *Contractor*'s share using the final Price for Work Done to Date and the final total of the Prices. This share is included in the final amount due.

53.5 If the *Project Manager* accepts a proposal by the *Contractor* to change the Works Information provided by the *Employer* which will reduce Actual Cost the Prices are not reduced.

The *bill of quantities* **55**

55.1 Information in the *bill of quantities* is not Works Information or Site Information

Compensation events **60**

60.4 A difference between the final total quantity of work done and a quantity for an item stated in the *bill of quantities* at the Contract Date is a compensation event if

- the difference causes the Actual Cost per unit of quantity to change and
- the rate in the *bill of quantities* for the item at the Contract Date multiplied by the final total quantity of work done is more than 0.1% of the total of the Prices at the Contract Date.

If the Actual Cost per unit of quantity is reduced, the affected rate is reduced.

60.5 A difference between the final total quantity of work done and the quantity for an item stated in the *bill of quantities* at the Contract Date which delays Completion is a compensation event.

60.6 The *Project Manager* corrects mistakes in the *bill of quantities* which are departures from the *method of measurement* or are due to ambiguities or inconsistencies. Each such correction is a compensation event which may lead to reduced Prices.

Defined Cost **52**

52.2 The *Contractor* keeps these records

- accounts of payments of Defined Cost,
- proof that the payments have been made,
- communications about and assessments of compensation events for Subcontractors and
- other records as stated in the Works Information.

52.3 The *Contractor* allows the *Project Manager* to inspect at any time within working hours the accounts and records which he is required to keep.

The *Contractor*'s share **53**

53.5 The *Project Manager* assesses the *Contractor*'s share of the difference between the Total of the Prices and the Price for Work Done to Date. The difference is divided into increments falling within each of the *share ranges*. The limits of a *share range* are the Price for Work Done to Date divided by the Total of the Prices, expressed as a percentage. The *Contractor*'s share equals the sum of the products of the increment within each *share range* and the corresponding *Contractor's share percentage*.

53.6 If the Price for Work Done to Date is less than the Total of the Prices, the *Contractor* is paid his share of the saving. If the Price for Work Done to Date is greater than the Total of the Prices, the *Contractor* pays his share of the excess.

53.7 The *Project Manager* makes a preliminary assessment of the *Contractor*'s share at Completion of the whole of the *works* using his forecasts of the final Price for Work Done to Date and the final Total of the Prices. This share is included in the amount due following Completion of the whole of the *works*.

53.8 The *Project Manager* makes a final assessment of the *Contractor*'s share using the final Price for Work Done to Date and the final Total of the Prices. This share is included in the final amount due.

The Bill of Quantities **55**

55.1 Information in the Bill of Quantities is not Works Information or Site Information.

Compensation events **60**

60.4 A difference between the final total quantity of work done and the quantity stated for an item in the Bill of Quantities is a compensation event if

- the difference does not result from a change to the Works Information,
- the difference causes the Defined Cost per unit of quantity to change and
- the rate in the Bill of Quantities for the item multiplied by the final total quantity of work done is more than 0.5% of the total of the Prices at the Contract Date.

If the Defined Cost per unit of quantity is reduced, the affected rate is reduced.

60.5 A difference between the final total quantity of work done and the quantity for an item stated in the Bill of Quantities which delays Completion or the meeting of the Condition stated for a Key Date is a compensation event.

60.6 The *Project Manager* corrects mistakes in the Bill of Quantities which are departures from the rules for item descriptions and for division of the work into items in the *method of measurement* or are due to ambiguities or inconsistencies. Each such correction is a compensation event which may lead to reduced Prices.

60.7 In assessing a compensation event which results from a correction of an inconsistency between the Bill of Quantities and another document, the *Contractor* is assumed to have taken the Bill of Quantities as correct.

Assessing compensation **63**
events 63.2 If the effect of a compensation event is to reduce the total Actual Cost and t
event is

- a change to the Works Information or
- a correction of an assumption stated by the *Project Manager* for assessin
an earlier compensation event,

the Prices are reduced.

63.9 Assessments for changed Prices for compensation events are in the form
changes to the *bill of quantities*. If the *Project Manager* and the *Contract
agree, rates and lump sums in the *bill of quantities* may be used as a basis f
assessment instead of Actual Cost and the resulting Fee.

63.11 If the *Project Manager* and the *Contractor* agree, the *Contractor* assesses a co
pensation event using the Shorter Schedule of Cost Components. The *Proje
Manager* may make his own assessments using the Shorter Schedule of Cc
Components.

Implementing compensation **65**
events 65.4 The *Project Manager* includes the changes to the Prices and the Completi
Date from the quotation which he has accepted or from his own assessment
his notification implementing a compensation event.

Payment on termination **97**
97.4 If there is a termination, the *Project Manager* assesses the *Contractor*'s sha
after he has certified termination. His assessment uses the Price for Work Do
to Date at termination and the total of the Prices for the work done befc
termination.

Assessing compensation events	63	
	63.11	If the effect of a compensation event is to reduce the total Defined Cost and the event is

- a change to the Works Information, other than a change to the Works Information provided by the *Employer* which the *Contractor* proposed and the *Project Manager* has accepted or
- a correction of an assumption stated by the *Project Manager* for assessing an earlier compensation event,

the Prices are reduced.

63.13 Assessments for changed Prices for compensation events are in the form of changes to the Bill of Quantities.

- For the whole or a part of a compensation event for work not yet done and for which there is an item in the Bill of Quantities, the changes are

  - a changed rate,
  - a changed quantity or
  - a changed lump sum.

- For the whole or a part of a compensation event for work not yet done and for which there is no item in the Bill of Quantities, the change is a new priced item which, unless the *Project Manager* and the *Contractor* agree otherwise, is compiled in accordance with the *method of measurement*.
- For the whole or a part of a compensation event for work already done, the change is a new lump sum item.

If the *Project Manager* and the *Contractor* agree, rates and lump sums may be used to assess a compensation event instead of Defined Cost.

63.15 If the *Project Manager* and the *Contractor* agree, the *Contractor* assesses a compensation event using the Shorter Schedule of Cost Components. The *Project Manager* may make his own assessments using the Shorter Schedule of Cost Components.

Implementing compensation events	65	
	65.4	The changes to the Prices, the Completion Date and the Key Dates are included in the notification implementing a compensation event.

Payment on termination	93	
	93.5	If there is a termination, the *Project Manager* assesses the *Contractor*'s share after he has certified termination. His assessment uses, as the Price for Work Done to Date, the total of the Defined Cost which the *Contractor* has paid and which he is committed to pay for work done before termination.
	93.6	The *Project Manager*'s assessment of the *Contractor*'s share is added to the amounts due to the *Contractor* on termination if there has been a saving or deducted if there has been an excess.

Option E: Cost reimbursable contract

Identified and defined terms   **11**

        11.2   (27) Actual Cost is the amount of payments due to Subcontractors for work which is subcontracted and the cost of the components in the Schedule of Cost Components for work which is not subcontracted, less any Disallowed Cost.

                (30) Disallowed Cost is cost which the *Project Manager* decides

- is not justified by the *Contractor*'s accounts and records,
- should not have been paid to a Subcontractor in accordance with his subcontract,
- was incurred only because the *Contractor* did not
  - follow an acceptance or procurement procedure stated in the Works Information or
  - give an early warning which he could have given or
- results from paying a Subcontractor more for a compensation event than is included in the accepted quotation or assessment for the compensation event

and the cost of

- correcting Defects after Completion,
- correcting Defects caused by the *Contractor* not complying with a requirement for how he is to Provide the Works stated in the Works Information,
- Plant and Materials not used to Provide the Works (after allowing for reasonable wastage) and
- resources not used to Provide the Works (after allowing for reasonable availability and utilisation) or not taken away from the Working Area when the *Project Manager* requested.

                (23) The Price for Work Done to Date is the Actual Cost which the *Contractor* has paid plus the Fee.

                (19) The Prices are the Actual Cost plus the Fee.

Providing the Works   **20**

        20.3   The *Contractor* advises the *Project Manager* on the practical implications of the design of the *works* and on subcontracting arrangements.

        20.4   The *Contractor* prepares forecasts of the total Actual Cost for the whole of the *works* in consultation with the *Project Manager* and submits them to the *Project Manager*. Forecasts are prepared at the intervals stated in the Contract Data from the *starting date* until Completion of the whole of the *works*. An explanation of the changes made since the previous forecast is submitted with each forecast.

ption E: Cost reimbursable contract

Identified and defined **11**
terms 11.2   (23) Defined Cost is

- the amount of payments due to Subcontractors for work which is subcontracted without taking account of amounts deducted for
  - retention,
  - payment to the *Employer* as a result of the Subcontractor failing to meet a Key Date,
  - the correction of Defects after Completion,
  - payments to Others and
  - the supply of equipment, supplies and services included in the charge for overhead cost within the Working Areas in this contract

and

- the cost of components in the Schedule of Cost Components for other work

less Disallowed Cost.

(25) Disallowed Cost is cost which the *Project Manager* decides

- is not justified by the *Contractor*'s accounts and records,
- should not have been paid to a Subcontractor or supplier in accordance with his contract,
- was incurred only because the *Contractor* did not
  - follow an acceptance or procurement procedure stated in the Works Information or
  - give an early warning which this contract required him to give

and the cost of

- correcting Defects after Completion,
- correcting Defects caused by the *Contractor* not complying with a constraint on how he is to Provide the Works stated in the Works Information,
- Plant and Materials not used to Provide the Works (after allowing for reasonable wastage) unless resulting from a change to the Works Information,
- resources not used to Provide the Works (after allowing for reasonable availability and utilisation) or not taken away from the Working Areas when the *Project Manager* requested and
- preparation for and conduct of an adjudication or proceedings of the *tribunal*.

(29) The Price for Work Done to Date is the total Defined Cost which the *Project Manager* forecasts will have been paid by the *Contractor* before the next assessment date plus the Fee.

(32) The Prices are the Defined Cost plus the Fee.

Providing the Works **20**
20.3   The *Contractor* advises the *Project Manager* on the practical implications of the design of the *works* and on subcontracting arrangements.

20.4   The *Contractor* prepares forecasts of the total Defined Cost for the whole of the *works* in consultation with the *Project Manager* and submits them to the *Project Manager*. Forecasts are prepared at the intervals stated in the Contract Data from the *starting date* until Completion of the whole of the *works*. An explanation of the changes made since the previous forecast is submitted with each forecast.

Subcontracting 26

26.4   The *Contractor* submits the proposed contract data for each subcontract f acceptance to the *Project Manager* if

- the NEC Engineering and Construction Subcontract or the NEC Profe sional Services Contract is to be used and

- the *Project Manager* instructs the *Contractor* to make the submission.

A reason for not accepting the proposed contract data is that its use will n allow the *Contractor* to Provide the Works.

Acceleration 36

36.4   When the *Project Manager* accepts a quotation for an acceleration, he chang the Completion Date accordingly and accepts the revised programme.

36.5   The *Contractor* submits a Subcontractor's proposal to accelerate to the *Proje Manager* for acceptance.

Assessing the amount due 50

50.7   Payments of Actual Cost made by the *Contractor* in a currency other than t *currency of this contract* are included in the amount due as payments to be ma to him in the same currency. Such payments are converted to the *currency this contract* in order to calculate the Fee at the *exchange rates*.

Actual Cost 52

52.2   The *Contractor* keeps

- accounts of his payments of Actual Cost,

- records which show that the payments have been made,

- records of communications and calculations relating to assessment of co pensation events for Subcontractors and

- other accounts and records as stated in the Works Information.

52.3   The *Contractor* allows the *Project Manager* to inspect at any time with working hours the accounts and records which he is required to keep.

Assessing compensation events 63

63.11   If the *Project Manager* and the *Contractor* agree, the *Contractor* assesses a co pensation event using the Shorter Schedule of Cost Components. The *Proje Manager* may make his own assessments using the Shorter Schedule of Cc Components.

Implementing compensation events 65

65.3   The *Project Manager* includes the changes to the forecast amount of the Pri and the Completion Date in his notification to the *Contractor* implementing compensation event.

65.5   The *Contractor* does not implement a subcontract compensation event until has been agreed by the *Project Manager*.

Subcontracting	**26**

26.4   The *Contractor* submits the proposed contract data for each subcontract for acceptance to the *Project Manager* if

- an NEC contract is proposed and
- the *Project Manager* instructs the *Contractor* to make the submission.

A reason for not accepting the proposed contract data is that its use will not allow the *Contractor* to Provide the Works.

Acceleration	**36**

36.4   When the *Project Manager* accepts a quotation for an acceleration, he changes the Completion Date, the Key Dates and the forecast of the total Defined Cost of the whole of the *works* accordingly and accepts the revised programme.

Tests and inspections	**40**

40.7   When the *Project Manager* assesses the cost incurred by the *Employer* in repeating a test or inspection after a Defect is found, the *Project Manager* does not include the *Contractor*'s cost of carrying out the repeat test or inspection.

Assessing the amount due	**50**

50.7   Payments of Defined Cost made by the *Contractor* in a currency other than the *currency of this contract* are included in the amount due as payments to be made to him in the same currency. Such payments are converted to the *currency of this contract* in order to calculate the Fee using the *exchange rates.*

Defined Cost	**52**

52.2   The *Contractor* keeps these records

- accounts of payments of Defined Cost,
- proof that the payments have been made,
- communications about and assessments of compensation events for Subcontractors and
- other records as stated in the Works Information.

52.3   The *Contractor* allows the *Project Manager* to inspect at any time within working hours the accounts and records which he is required to keep.

Assessing compensation events	**63**

63.15   If the *Project Manager* and the *Contractor* agree, the *Contractor* assesses a compensation event using the Shorter Schedule of Cost Components. The *Project Manager* may make his own assessments using the Shorter Schedule of Cost Components.

Implementing compensation events	**65**

65.3   The changes to the forecast amount of the Prices, the Completion Date and the Key Dates are included in the notification implementing a compensation event.

Option F: Management contract

Identified and defined terms 11

11.2 (26) Actual Cost is the amount of payments due to Subcontractors for wor which the *Contractor* is required to subcontract, less any Disallowed Cost.

(29) Disallowed Cost is cost which the *Project Manager* decides

- is not justified by the accounts and records provided by the *Contractor*,
- should not have been paid to a Subcontractor in accordance with hi subcontract,
- was incurred only because the *Contractor* did not
  - follow an acceptance or procurement procedure stated in the Work Information or
  - give an early warning which he could have given or
- results from paying a Subcontractor more for a compensation event than i included in the accepted quotation or assessment for the compensatio event.

(22) The Price for Work Done to Date is the amount of Actual Cost which th *Contractor* has accepted for payment plus the Fee.

(19) The Prices are the Actual Cost plus the Fee.

Providing the Works 20

20.2 The *Contractor* manages the *Contractor*'s design and the construction and insta lation of the works. The *Contractor* subcontracts design, construction and insta lation of the *works* and other work which is stated in the Works Information a to be subcontracted. He either does other work which is not stated in the Work Information as to be subcontracted himself or subcontracts it.

Option F: Management contract

Identified and defined terms

11

11.2

(24) Defined Cost is

- the amount of payments due to Subcontractors for work which is subcontracted without taking account of amounts deducted for

  - retention,
  - payment to the *Employer* as a result of the Subcontractor failing to meet a Key Date,
  - the correction of Defects after Completion,
  - payments to Others,
  - the supply of equipment, supplies and services included in the charge for overhead cost within the Working Areas in this contract

and

- the *prices* for work done by the *Contractor* himself

less Disallowed Cost.

> This main Option now includes provision for paying the *Contractor* a lump sum via '*Prices*' for the work done by the *Contractor* himself that is listed in the Contract Data. In ECC2, the *Contractor* recovered such costs through the Fee. This is a more transparent basis and should reduce the amount of Fee and some uncertainty.

(26) Disallowed Cost is cost which the *Project Manager* decides

- is not justified by the accounts and records provided by the *Contractor*,
- should not have been paid to a Subcontractor or supplier in accordance with his contract,
- was incurred only because the *Contractor* did not

  - follow an acceptance or procurement procedure stated in the Works Information or
  - give an early warning which this contract required him to give or

- is a payment to a Subcontractor for

  - work which the Contract Data states that the *Contractor* will do himself or
  - the *Contractor*'s management.

> The list of items of Disallowed Cost is extended to include any payments to Subcontractors for works which the Contract Data states that the *Contractor* will do himself and also for payments to Subcontractors for *Contractor*'s management. The *Contractor* may choose to alter the amount of work he does himself or how he manages it as the contract proceeds but he recovers payment for these aspects on the basis tendered in the Contract Data.

(29) The Price for Work Done to Date is the total Defined Cost which the *Project Manager* forecasts will have been paid by the *Contractor* before the next assessment date plus the Fee.

(32) The Prices are the Defined Cost plus the Fee.

Providing the Works

20

20.2

The *Contractor* manages the *Contractor*'s design, the provision of Site services and the construction and installation of the *works*. The *Contractor* subcontracts the *Contractor*'s design, the provision of Site services and the construction and installation of the *works* except work which the Contract Data states that he will do himself.

| | 20.3 | The *Contractor* advises the *Project Manager* on the practical implications of th design of the *works* and on subcontracting arrangements. |

20.3   The *Contractor* advises the *Project Manager* on the practical implications of th design of the *works* and on subcontracting arrangements.

20.4   The *Contractor* prepares forecasts of the total Actual Cost for the whole of th *works* in consultation with the *Project Manager* and submits them to the *Proje Manager*. Forecasts are prepared at the intervals stated in the Contract Da¹ from the *starting date* until Completion of the whole of the *works*. An explana tion of the changes made since the previous forecast is submitted with eac forecast.

Subcontracting   26

26.4   The *Contractor* submits the proposed contract data for each subcontract f acceptance to the *Project Manager* if

- the NEC Engineering and Construction Subcontract or the NEC Profe sional Services Contract is to be used and

- the *Project Manager* instructs the *Contractor* to make the submission.

A reason for not accepting the proposed contract data is that its use will n allow the *Contractor* to Provide the Works.

Acceleration   36

36.4   When the *Project Manager* accepts a quotation for an acceleration, he chang the Completion Date accordingly and accepts the revised programme.

36.5   The *Contractor* submits a Subcontractor's proposal to accelerate to the *Proje Manager* for acceptance.

Assessing the amount due   50

50.7   Payments of Actual Cost made by the *Contractor* in a currency other than th *currency of this contract* are included in the amount due as payments to be mad to him in the same currency. Such payments are converted to the *currency this contract* in order to calculate the Fee at the *exchange rates*.

Actual Cost   52

52.2   The *Contractor* keeps

- accounts of his payments of Actual Cost,

- records which show that the payments have been made,

- records of communications and calculations relating to assessment compensation events for Subcontractors and

- other accounts and records as stated in the Works Information.

52.3   The *Contractor* allows the *Project Manager* to inspect at any time withi working hours the accounts and records which he is required to keep.

Implementing compensation   65
events   65.3   The *Project Manager* includes the changes to the forecast amount of the Price and the Completion Date in his notification to the *Contractor* implementing compensation event.

65.5   The *Contractor* does not implement a subcontract compensation event until has been agreed by the *Project Manager*.

20.3   The *Contractor* advises the *Project Manager* on the practical implications of the design of the *works* and on subcontracting arrangements.

20.4   The *Contractor* prepares forecasts of the total Defined Cost for the whole of the *works* in consultation with the *Project Manager* and submits them to the *Project Manager*. Forecasts are prepared at the intervals stated in the Contract Data from the *starting date* until Completion of the whole of the *works*. An explanation of the changes made since the previous forecast is submitted with each forecast.

20.5   If work which the *Contractor* is to do himself is affected by a compensation event, the *Project Manager* and the *Contractor* agree the change to the price for the work and any change to the Completion Date and Key Dates. If they cannot agree, the *Project Manager* decides the change.

> This new subclause details the process for agreeing any change to the *price*, Completion Date and Key Date in the event that any compensation event affects the work which the Contract Data states the *Contractor* is to do himself.

Subcontracting **26**
26.4   The *Contractor* submits the proposed contract data for each subcontract for acceptance to the *Project Manager* if

- an NEC contract is proposed and
- the *Project Manager* instructs the *Contractor* to make the submission.

A reason for not accepting the proposed contract data is that its use will not allow the *Contractor* to Provide the Works.

Acceleration **36**
36.4   When the *Project Manager* accepts a quotation for an acceleration, he changes the Completion Date, the Key Dates and the forecast of the total Defined Cost of the whole of the *works* accordingly and accepts the revised programme.

Assessing the amount due **50**
50.7   Payments of Defined Cost made by the *Contractor* in a currency other than the *currency of this contract* are included in the amount due as payments to be made to him in the same currency. Such payments are converted to the *currency of this contract* in order to calculate the Fee using the *exchange rates*.

Defined Cost **52**
52.2   The *Contractor* keeps these records

- accounts of payments of Defined Cost,
- proof that the payments have been made,
- communications about and assessments of compensation events for Subcontractors and
- other records as stated in the Works Information.

52.3   The *Contractor* allows the *Project Manager* to inspect at any time within working hours the accounts and records which he is required to keep.

Implementing compensation events **65**
65.3   The changes to the forecast amount of the Prices, the Completion Date and the Key Dates are included in the notification implementing a compensation event.

9 Disputes and termination

Settlement of disputes **90**

90.1 Any dispute arising under or in connection with this contract is submitted to and settled by the *Adjudicator* as follows.

The *Adjudicator* **92**

92.1 The *Adjudicator* settles the dispute as independent adjudicator and not as arbitrator.

92.2 If the *Adjudicator* resigns or is unable to act, the Parties choose a new adjudicator jointly. If the Parties have not chosen a new adjudicator jointly within four weeks of the *Adjudicator* resigning or becoming unable to act, a Party may ask the person stated in the Contract Data to choose a new adjudicator and the Parties accept his choice. The new adjudicator is appointed as *Adjudicator* under the NEC Adjudicator's Contract.

92.2 He has power to settle disputes that were currently submitted to his predecessor but had not been settled at the time when his predecessor resigned or became unable to act. The date of his appointment is the date of submission of these disputes to him as *Adjudicator*.

DISPUTE RESOLUTION

Option W1

A completely amended procedure is provided in ECC3 for resolving disputes. Should the United Kingdom Housing Grants, Construction and Regeneration Act 1996 (the 'Act') not apply it is anticipated the Parties will use Option W1. Option W1 is the standard NEC approach to adjudication and is not act-compliant. Option W1 in this commentary is compared with ECC2 clauses 90 to 93.

Dispute resolution procedure (used unless the United Kingdom Housing Grants, Construction and Regeneration Act 1996 applies).

Dispute resolution	**W1**	
	W1.1	A dispute arising under or in connection with this contract is referred to and decided by the *Adjudicator*.

The *Adjudicator* is now deciding the dispute, not settling it as in ECC2.

The *Adjudicator*	W1.2	(1) The Parties appoint the *Adjudicator* under the NEC Adjudicator's Contract current at the *starting date*.

The *Adjudicator* is to be appointed using the latest version of the NEC Adjudicator's Contract.

(2) The *Adjudicator* acts impartially and decides the dispute as an independent adjudicator and not as an arbitrator.

(3) If the *Adjudicator* is not identified in the Contract Data or if the *Adjudicator* resigns or is unable to act, the Parties choose a new adjudicator jointly. If the Parties have not chosen an adjudicator, either Party may ask the *Adjudicator nominating body* to choose one. The *Adjudicator nominating body* chooses an adjudicator within four days of the request. The chosen adjudicator becomes the *Adjudicator*.

It is not mandatory to identify the *Adjudicator* in the Contract Data. Where one is not named, or the *Adjudicator* resigns or is unable to act, then the Parties are to jointly choose a new *Adjudicator*. If they do not choose one then a new provision allows either Party to ask the *Adjudicator nominating body* to choose an *Adjudicator*. It follows, along with a number of similar provisions in NEC contracts, that the Contract Data requires comprehensive drafting at tender stage.

(4) A replacement *Adjudicator* has the power to decide a dispute referred to his predecessor but not decided at the time when the predecessor resigned or became unable to act. He deals with an undecided dispute as if it had been referred to him on the date he was appointed.

(5) The *Adjudicator*, his employees and agents are not liable to the Parties for any action or failure to take action in an adjudication unless the action or failure to take action was in bad faith.

This subclause sets out the basis of liability the *Adjudicator* has to the Parties.

Settlement of disputes 90.1

ADJUDICATION TABLE

Dispute about:	Which Party may submit it to the *Adjudicator*?	When may it be submitted to the *Adjudicator*?
An action of the *Project Manager* or the *Supervisor*	The *Contractor*	Between two and four weeks after the *Contractor*'s notification of the dispute to the *Project Manager*, the notification itself being made not more than four weeks after the *Contractor* becomes aware of the action
The *Project Manager* or *Supervisor* not having taken an action	The *Contractor*	Between two and four weeks after the *Contractor*'s notification of the dispute to the *Project Manager*, the notification itself being made not more than four weeks after the *Contractor* becomes aware that the action was not taken
Any other matter	Either Party	Between two and four weeks after notification of the dispute to the other Party and the *Project Manager*

The adjudication **91**

91.1   The Party submitting the dispute to the *Adjudicator* includes with his submission information to be considered by the *Adjudicator*. Any further information from a Party to be considered by the *Adjudicator* is provided within four weeks from the submission.

The adjudication   W1.3   (1) Disputes are notified and referred to the *Adjudicator* in accordance with the Adjudication Table.

ADJUDICATION TABLE

Dispute about	Which Party may refer it to the *Adjudicator*?	When may it be referred to the *Adjudicator*?
An action of the *Project Manager* or the *Supervisor*	The *Contractor*	Between two and four weeks after the *Contractor*'s notification of the dispute to the *Employer* and the *Project Manager*, the notification itself being made not more than four weeks after the *Contractor* becomes aware of the action
The *Project Manager* or *Supervisor* not having taken an action	The *Contractor*	Between two and four weeks after the *Contractor*'s notification of the dispute to the *Employer* and the *Project Manager*, the notification itself being made not more than four weeks after the *Contractor* becomes aware that the action was not taken
A quotation for a compensation event which is treated as having been accepted	The *Employer*	Between two and four weeks after the *Project Manager*'s notification of the dispute to the *Employer* and the *Contractor*, the notification itself being made not more than four weeks after the quotation was treated as accepted
Any other matter	Either Party	Between two and four weeks after notification of the dispute to the other Party and the *Project Manager*

A new provision included in the Adjudication Table is that the *Employer* may bring about adjudication on the new ECC3 provisions for quotations for compensation events having been treated as accepted.

(2) The times for notifying and referring a dispute may be extended by the *Project Manager* if the *Contractor* and the *Project Manager* agree to the extension before the notice or referral is due. The *Project Manager* notifies the extension that has been agreed to the *Contractor*. If a disputed matter is not notified and referred within the times set out in this contract, neither Party may subsequently refer it to the *Adjudicator* or the *tribunal*.

(3) The Party referring the dispute to the *Adjudicator* includes with his referral information to be considered by the *Adjudicator*. Any more information from a Party to be considered by the *Adjudicator* is provided within four weeks of the referral. This period may be extended if the *Adjudicator* and the Parties agree.

91.2 If a matter disputed under or in connection with a subcontract is also a matter disputed under or in connection with this contract, the *Contractor* may submit the subcontract dispute to the *Adjudicator* at the same time as the main contract submission. The *Adjudicator* then settles the two disputes together and references to the Parties for the purposes of the dispute are interpreted as including the Subcontractor.

92.1 The *Adjudicator*'s powers include the power to review and revise any action or inaction of the *Project Manager* or *Supervisor* related to the dispute.

92.1 Any communication between a Party and the *Adjudicator* is communicated also to the other Party.

92.1 If the *Adjudicator*'s decision includes assessment of additional cost or delay caused to the *Contractor*, he makes his assessment in the same way as a compensation event is assessed.

91.1 The *Adjudicator* notifies his decision within four weeks of the end of the period for providing information. The four week periods in this clause may be extended if requested by the *Adjudicator* in view of the nature of the dispute and agreed by the Parties.

90.2 The *Adjudicator* settles the dispute by notifying the Parties and the *Project Manager* of his decision together with his reasons within the time allowed by this contract.

90.2 Unless and until there is such a settlement, the Parties and the *Project Manager* proceed as if the action, inaction or other matter disputed were not disputed.

90.2 The decision is final and binding unless and until revised by the *tribunal*.

92.1 His decision is enforceable as a matter of contractual obligation between the Parties and not as an arbitral award.

(4) If a matter disputed by the *Contractor* under or in connection with a subcontract is also a matter disputed under or in connection with this contract and if the subcontract allows, the *Contractor* may refer the subcontract dispute to the *Adjudicator* at the same time as the main contract referral. The *Adjudicator* then decides the disputes together and references to the Parties for the purposes of the dispute are interpreted as including the Subcontractor.

The subcontract conditions must allow for combining disputes to permit multi-party adjudications.

(5) The *Adjudicator* may

- review and revise any action or inaction of the *Project Manager* or *Supervisor* related to the dispute and alter a quotation which has been treated as having been accepted,
- take the initiative in ascertaining the facts and the law related to the dispute,
- instruct a Party to provide further information related to the dispute within a stated time and
- instruct a Party to take any other action which he considers necessary to reach his decision and to do so within a stated time.

The *Adjudicator* can take the initiative in ascertaining the facts and the law related to the dispute should he choose to do so. A more pro-active position for the *Adjudicator* is therefore encouraged.

(6) A communication between a Party and the *Adjudicator* is communicated to the other Party at the same time.

(7) If the *Adjudicator*'s decision includes assessment of additional cost or delay caused to the *Contractor*, he makes his assessment in the same way as a compensation event is assessed.

(8) The *Adjudicator* decides the dispute and notifies the Parties and the *Project Manager* of his decision and his reasons within four weeks of the end of the period for receiving information. This four week period may be extended if the Parties agree.

The *Adjudicator* must give both his decision and reasons within the stated time or an extended time if both Parties agree.

(9) Unless and until the *Adjudicator* has notified the Parties of his decision, the Parties, the *Project Manager* and the *Supervisor* proceed as if the matter disputed was not disputed.

(10) The *Adjudicator*'s decision is binding on the Parties unless and until revised by the *tribunal* and is enforceable as a matter of contractual obligation between the Parties and not as an arbitral award. The *Adjudicator*'s decision is final and binding if neither Party has notified the other within the times required by this contract that he is dissatisfied with a decision of the *Adjudicator* and intends to refer the matter to the *tribunal*.

Either Party can refer the *Adjudicator*'s decision to the *tribunal* but must notify to the other Party such dissatisfaction with the decision and an intention to refer the matter to the *tribunal*, again within the stated timescales. Failure to do this renders the *Adjudicator*'s decision final and binding.

(11) The *Adjudicator* may, within two weeks of giving his decision to the Parties, correct any clerical mistake or ambiguity.

This new subclause results in the *Adjudicator* having a limited time period to correct a clerical mistake or ambiguity, which is a practical measure.

Review by the *tribunal* **93**

93.1 If after the *Adjudicator*

- notifies his decision or

- fails to do so

within the time provided by this contract a Party is dissatisfied, that Party notifies the other Party of his intention to refer the matter which he disputes to the *tribunal*. It is not referable to the *tribunal* unless the dissatisfied Party notifies his intention within four weeks of

- notification of the *Adjudicator*'s decision or

- the time provided by this contract for this notification if the *Adjudicator* fails to notify his decision within that time

whichever is the earlier. The *tribunal* proceedings are not started before Completion of the whole of the *works* or earlier termination.

93.2 The *tribunal* settles the dispute referred to it. Its powers include the power to review and revise any decision of the *Adjudicator* and any action or inaction of the *Project Manager* or the *Supervisor* related to the dispute. A Party is not limited in the *tribunal* proceedings to the information, evidence or arguments put to the *Adjudicator*.

Review by the *tribunal*   W1.4   (1) A Party does not refer any dispute under or in connection with this contract to the *tribunal* unless it has first been referred to the *Adjudicator* in accordance with this contract.

Any dispute must be referred to an *Adjudicator* before it can go to the *tribunal*.

(2) If, after the *Adjudicator* notifies his decision a Party is dissatisfied, he may notify the other Party that he intends to refer it to the *tribunal*. A Party may not refer a dispute to the *tribunal* unless this notification is given within four weeks of notification of the *Adjudicator*'s decision.

(3) If the *Adjudicator* does not notify his decision within the time provided by this contract, a Party may notify the other Party that he intends to refer the dispute to the *tribunal*. A Party may not refer a dispute to the *tribunal* unless this notification is given within four weeks of the date by which the *Adjudicator* should have notified his decision.

This new subclause deals with the instance when an *Adjudicator* does not notify his decision within the stated time limits. In this event, a Party may notify the other Party that he intends to refer the dispute to the *tribunal*.

(4) The *tribunal* settles the dispute referred to it. The *tribunal* has the powers to reconsider any decision of the *Adjudicator* and review and revise any action or inaction of the *Project Manager* or the *Supervisor* related to the dispute. A Party is not limited in the *tribunal* proceedings to the information, evidence or arguments put to the *Adjudicator*.

(5) If the *tribunal* is arbitration, the *arbitration procedure*, the place where the arbitration is to be held and the method of choosing the arbitrator are those stated in the Contract Data.

(6) A Party does not call the *Adjudicator* as a witness in *tribunal* proceedings.

The *Adjudicator* cannot be subsequently called as a witness in any *tribunal* proceedings.

9 Disputes and termination

The following clauses incorporate Addendum Y(UK)2 Contract (Ref NEC/ECC/Y(UK)2/April 1998) to take into accou
The Housing Grants, Construction and Regeneration Act 1996 (Part II).

Avoidance and settlement of **90**
disputes 90.1 The Parties and the *Project Manager* follow this procedure for the avoidan
and settlement of disputes.

90.2 If the *Contractor* is dissatisfied with an action or a failure to take action by th
Project Manager, he notifies his dissatisfaction to the *Project Manager* no late
than

- four weeks after he became aware of the action or
- four weeks after he became aware that the action had not been taken.

Within two weeks of such notification of dissatisfaction, the *Contractor* and th
Project Manager attend a meeting to discuss and seek to resolve the matter.

90.3 If either Party is dissatisfied with any other matter, he notifies his dissatisfactio
to the *Project Manager* and to the other Party no later than four weeks after l
became aware of the matter. Within two weeks of such notification of dissati:
faction, the Parties and the *Project Manager* attend a meeting to discuss an
seek to resolve the matter.

90.4 The Parties agree that no matter shall be a dispute unless a notice of dissatisfa
tion has been given and the matter has not been resolved within four weeks. Th
word dispute (which includes a difference) has that meaning.

90.8 The *Adjudicator* acts impartially.

The *Adjudicator* 92.1 The *Adjudicator* gives his decision on the dispute as independent adjudicato
and not as arbitrator.

92.2 If the *Adjudicator* resigns or is unable to act, the Parties choose a new adjudica
tor jointly. If the Parties have not chosen a new adjudicator jointly within fou
weeks of the *Adjudicator* resigning or becoming unable to act, a Party may as
the person stated in the Contract Data to choose a new adjudicator and th
Parties accept his choice. The new adjudicator is appointed as *Adjudicator* unde
the NEC Adjudicator's Contract. He has power to decide on disputes that wer
currently submitted to his predecessor but a decision had not been given at th
time when his predecessor resigned or became unable to act. The date of h
appointment is the date of submission of these disputes to him as *Adjudicator*.

90.12 The *Adjudicator* is not liable for anything done or omitted in the discharge c
purported discharge of his functions as adjudicator unless the act or omission i
in bad faith and any employee or agent of the *Adjudicator* is similarly protecte
from liability.

90.5 Either Party may give notice to the other Party at any time of his intention t
refer a dispute to adjudication. The notifying Party refers the dispute to th
Adjudicator within seven days of the notice.

90.6 The Party referring the dispute to the *Adjudicator* includes with his submissio
information to be considered by the *Adjudicator*. Any further information fror
a Party to be considered by the *Adjudicator* is provided within fourteen days c
referral.

ption W2

completely amended procedure is provided in ECC3 for resolving disputes. Should the United Kingdom Housing ▌rants, Construction and Regeneration Act 1996 apply (the 'Act') it is anticipated the Parties will use Option W2. ▌ption W2 is the Act-compliant NEC approach to adjudication. Option W2 in this commentary is compared with ▌CC2 clauses 90 to 93 as amended by Addendum Y(UK)2 Contract (Ref NEC/ECC/Y(UK)2/April 1998). These ▌auses are consolidated and reproduced in full with the ECC2 clauses. Any clauses that are similar between ▌ptions W1 and W2 are not commented on again below. The ECC2 notice of dissatisfaction provisions have been ▌eleted in W2.

▌spute resolution procedure (used in the United Kingdom when the Housing Grants, Construction and Regeneration ▌t 1996 applies).

Dispute resolution	**W2**	
	W2.1	(1) A dispute arising under or in connection with this contract is referred to and decided by the *Adjudicator*. A Party may refer a dispute to the *Adjudicator* at any time.

A feature of the Act is that a dispute can be referred at any time to adjudication.

(2) In this Option, time periods stated in days exclude Christmas Day, Good Friday and bank holidays.

The *Adjudicator*   W2.2   (1) The Parties appoint the *Adjudicator* under the NEC Adjudicator's Contract current at the *starting date*.

(2) The *Adjudicator* acts impartially and decides the dispute as an independent adjudicator and not as an arbitrator.

(3) If the *Adjudicator* is not identified in the Contract Data or if the *Adjudicator* resigns or becomes unable to act

- the Parties may choose an adjudicator jointly or
- a Party may ask the *Adjudicator nominating body* to choose an adjudicator.

The *Adjudicator nominating body* chooses an adjudicator within four days of the request. The chosen adjudicator becomes the *Adjudicator*.

(4) A replacement *Adjudicator* has the power to decide a dispute referred to his predecessor but not decided at the time when his predecessor resigned or became unable to act. He deals with an undecided dispute as if it had been referred to him on the date he was appointed.

(5) The *Adjudicator*, his employees and agents are not liable to the Parties for any action or failure to take action in an adjudication unless the action or failure to take action was in bad faith.

Combining procedures 91.1 If a matter causing dissatisfaction under or in connection with a subcontract also a matter causing dissatisfaction under or in connection with this contrac the subcontractor may attend the meeting between the Parties and the *Proje Manager* to discuss and seek to resolve the matter.

91.2 If a matter disputed under or in connection with a subcontract is also a matte disputed under or in connection with this contract, the *Contractor* may subm the subcontract dispute to the *Adjudicator* at the same time as the main contra submission. The *Adjudicator* then gives his decision on the two disputes togethe and references to the Parties for the purposes of the dispute are interpreted a including the Subcontractor.

92.1 The *Adjudicator*'s powers include the power to review and revise any action c inaction of the *Project Manager* or *Supervisor* related to the dispute.

90.8 The *Adjudicator* may take the initiative in ascertaining the facts and the law.

92.1 Any communication between a Party and the *Adjudicator* is communicated als to the other Party.

The adjudication W2.3 (1) Before a Party refers a dispute to the *Adjudicator*, he gives a notice of adjudication to the other Party with a brief description of the dispute and the decision which he wishes the *Adjudicator* to make. If the *Adjudicator* is named in the Contract Data, the Party sends a copy of the notice of adjudication to the *Adjudicator* when it is issued. Within three days of the receipt of the notice of adjudication, the *Adjudicator* notifies the Parties

- that he is able to decide the dispute in accordance with the contract or
- that he is unable to decide the dispute and has resigned.

If the *Adjudicator* does not so notify within three days of the issue of the notice of adjudication, either Party may act as if he has resigned.

This new subclause alerts the *Adjudicator* early giving pertinent information before the formal submission of the detail takes place. Following this and due to the very tight timescales for deciding disputes in adjudication provided under the Act, the *Adjudicator* notifies the Parties that he is or is not able to decide the dispute within 3 days of the issue of the notice of adjudication. If this notification does not so notify within the stated time, either Party may act as if the *Adjudicator* has resigned and the process effectively starts again with a new adjudicator.

(2) Within seven days of a Party giving a notice of adjudication he

- refers the dispute to the *Adjudicator*,
- provides the *Adjudicator* with the information on which he relies, including any supporting documents and
- provides a copy of the information and supporting documents he has provided to the *Adjudicator* to the other Party.

Any further information from a Party to be considered by the *Adjudicator* is provided within fourteen days of the referral. This period may be extended if the *Adjudicator* and the Parties agree.

(3) If a matter disputed by the *Contractor* under or in connection with a subcontract is also a matter disputed under or in connection with this contract, the *Contractor* may, with the consent of the Subcontractor, refer the subcontract dispute to the *Adjudicator* at the same time as the main contract referral. The *Adjudicator* then decides the disputes together and references to the Parties for the purposes of the dispute are interpreted as including the Subcontractor.

Any disputed matter which is also a matter disputed in a subcontract can also be referred to the *Adjudicator* at the same time, but only with consent of the Subcontractor.

(4) The *Adjudicator* may

- review and revise any action or inaction of the *Project Manager* or *Supervisor* related to the dispute and alter a quotation which has been treated as having been accepted,
- take the initiative in ascertaining the facts and the law related to the dispute,
- instruct a Party to provide further information related to the dispute within a stated time and
- instruct a Party to take any other action which he considers necessary to reach his decision and to do so within a stated time.

(5) If a Party does not comply with any instruction within the time stated by the *Adjudicator*, the *Adjudicator* may continue the adjudication and make his decision based upon the information and evidence he has received.

If a Party does not comply with any *Adjudicator*'s instruction, the *Adjudicator* may continue the adjudication, basing his decision on the information and evidence he has received.

92.1 If the *Adjudicator*'s decision includes assessment of additional cost or de caused to the *Contractor*, he makes his assessment in the same way as a compe sation event is assessed.

90.9 The *Adjudicator* reaches a decision within twenty eight days of referral or su longer period as is agreed by the Parties after the dispute has been referred. T *Adjudicator* may extend the period of twenty eight days by up to fourteen da with the consent of the notifying Party.

90.10 The *Adjudicator* provides his reasons to the Parties and to the *Project Manag* with his decision.

90.7 Unless and until the *Adjudicator* has given his decision on the dispute, t Parties and the *Project Manager* proceed as if the action, failure to take acti or other matters were not disputed.

90.11 The decision of the *Adjudicator* is binding until the dispute is finally determin by the *tribunal* or by agreement.

92.1 His decision is enforceable as a matter of contractual obligation between t Parties and not as an arbitral award.

Review by the *tribunal* **93**

93.1 If after the *Adjudicator*

- notifies his decision or

- fails to do so

within the time provided by this contract a Party is dissatisfied, that Par notifies the other Party of his intention to refer the matter which he disputes the *tribunal*. It is not referable to the *tribunal* unless the dissatisfied Party notif his intention within four weeks of

- notification of the *Adjudicator*'s decision or

- the time provided by this contract for this notification if the *Adjudicar* fails to notify his decision within that time

whichever is the earlier. The *tribunal* proceedings are not started before Comp tion of the whole of the *works* or earlier termination.

93.2 The *tribunal* settles the dispute referred to it. Its powers include the power review and revise any decision of the *Adjudicator* and any action or inaction the *Project Manager* or the *Supervisor* related to the dispute. A Party is m limited in the *tribunal* proceedings to the information, evidence or argumer put to the *Adjudicator*.

(6) A communication between a Party and the *Adjudicator* is communicated to the other Party at the same time.

(7) If the *Adjudicator's* decision includes assessment of additional cost or delay caused to the *Contractor,* he makes his assessment in the same way as a compensation event is assessed.

(8) The *Adjudicator* decides the dispute and notifies the Parties and the *Project Manager* of his decision and his reasons within twenty-eight days of the dispute being referred to him. This period may be extended by up to fourteen days with the consent of the referring Party or by any other period agreed by the Parties.

(9) Unless and until the *Adjudicator* has notified the Parties of his decision, the Parties, the *Project Manager* and the *Supervisor* proceed as if the matter disputed was not disputed.

(10) If the *Adjudicator* does not make his decision and notify it to the Parties within the time provided by this contract, the Parties and the *Adjudicator* may agree to extend the period for making his decision. If they do not agree to an extension, either Party may act as if the *Adjudicator* has resigned.

In the event that the *Adjudicator* fails to notify his decision to the Parties within the stated time, either the Parties agree to extend this period or one Party may act as if the *Adjudicator* has resigned.

(11) The *Adjudicator's* decision is binding on the Parties unless and until revised by the *tribunal* and is enforceable as a matter of contractual obligation between the Parties and not as an arbitral award. The *Adjudicator's* decision is final and binding if neither Party has notified the other within the times required by this contract that he is dissatisfied with a matter decided by the *Adjudicator* and intends to refer the matter to the *tribunal.*

(12) The *Adjudicator* may, within fourteen days of giving his decision to the Parties, correct a clerical mistake or ambiguity.

Review by the *tribunal* W2.4   (1) A Party does not refer any dispute under or in connection with this contract to the *tribunal* unless it has first been decided by the *Adjudicator* in accordance with this contract.

(2) If, after the *Adjudicator* notifies his decision a Party is dissatisfied, that Party may notify the other Party of the matter which he disputes and state that he intends to refer it to the *tribunal.* The dispute may not be referred to the *tribunal* unless this notification is given within four weeks of the notification of the *Adjudicator's* decision.

(3) The *tribunal* settles the dispute referred to it. The *tribunal* has the powers to reconsider any decision of the *Adjudicator* and to review and revise any action or inaction of the *Project Manager* or the *Supervisor* related to the dispute. A Party is not limited in *tribunal* proceedings to the information or evidence put to the *Adjudicator.*

(4) If the *tribunal* is arbitration, the *arbitration procedure,* the place where the arbitration is to be held and the method of choosing the arbitrator are those stated in the Contract Data.

(5) A Party does not call the *Adjudicator* as a witness in *tribunal* proceedings.

SECONDARY OPTION CLAUSES

Option N: Price adjustment for inflation (used only with Options A, B, C and D)

Defined terms **N1**

N1.1 (a) The Base Date Index (B) is the latest available index before the *base date*.

The Latest Index (L) is the latest available index before the date of assessme of an amount due.

(c) The Price Adjustment Factor is the total of the products of each of t proportions stated in the Contract Data multiplied by $(L-B)/B$ for the ind linked to it.

Price Adjustment Factors **N2**

N2.1 If an index is changed after it has been used in calculating a Price Adjustme Factor, the calculation is repeated and a correction included in the next asse ment of the amount due.

N2.2 The Price Adjustment Factor calculated at the Completion Date for the who of the *works* is used for calculating price adjustment after this date.

Compensation events **N3**

N3.1 The Actual Cost for compensation events is assessed using the

- Actual Costs current at the time of assessing the compensation eve adjusted to *base date* by dividing by one plus the Price Adjustment Fact for the last assessment of the amount due and

- Actual Costs at *base date* levels for amounts calculated from rates stated the Contract Data for employees and Equipment.

Price adjustment Options **N4**
A and B

N4.1 Each amount due includes an amount for price adjustment which is the sum o

- the change in the Price for Work Done to Date since the last assessment the amount due multiplied by the Price Adjustment Factor for the date the current assessment,

- the amount for price adjustment included in the previous amount due an

- correcting amounts, not included elsewhere, which arise from changes indices used for assessing previous amounts for price adjustment.

Options C and D **N4.2** Each time the amount due is assessed, an amount for price adjustment is add to the total of the Prices which is the sum of

- the change in the Price for Work Done to Date since the last assessment the amount due multiplied by $(1-1/(1+PAF))$ where PAF is the Pr Adjustment Factor for the date of the current assessment and

- correcting amounts, not included elsewhere, which arise from changes indices used for assessing previous amounts for price adjustment.

SECONDARY OPTION CLAUSES

The clause numbering for the secondary Option clauses has been amended, replacing the G to Z style with X1, X2, retaining Z for additional conditions of contract and continuing the 'Y(UK)...' approach denoting national legislation. This could be extended to other than UK matters.

Option X1: Price adjustment for inflation (used only with Options A, B, C and D)

Defined terms	X1	
	X1.1	(a) The Base Date Index (B) is the latest available index before the *base date*.
		(b) The Latest Index (L) is the latest available index before the date of assessment of an amount due.
		(c) The Price Adjustment Factor is the total of the products of each of the proportions stated in the Contract Data multiplied by $(L - B)/B$ for the index linked to it.
Price Adjustment Factor	X1.2	If an index is changed after it has been used in calculating a Price Adjustment Factor, the calculation is repeated and a correction included in the next assessment of the amount due.
		The Price Adjustment Factor calculated at the Completion Date for the whole of the *works* is used for calculating price adjustment after this date.
Compensation events	X1.3	The Defined Cost for compensation events is assessed using the

- Defined Cost current at the time of assessing the compensation event adjusted to *base date* by dividing by one plus the Price Adjustment Factor for the last assessment of the amount due and
- Defined Cost at *base date* levels for amounts calculated from rates stated in the Contract Data for employees and Equipment.

Price adjustment Options A and B   X1.4   Each amount due includes an amount for price adjustment which is the sum of

- the change in the Price for Work Done to Date since the last assessment of the amount due multiplied by the Price Adjustment Factor for the date of the current assessment,
- the amount for price adjustment included in the previous amount due and
- correcting amounts, not included elsewhere, which arise from changes to indices used for assessing previous amounts for price adjustment.

Price adjustment Options C and D   X1.5   Each time the amount due is assessed, an amount for price adjustment is added to the total of the Prices which is the sum of

- the change in the Price for Work Done to Date since the last assessment of the amount due multiplied by $(PAF/(1 + PAF))$ where PAF is the Price Adjustment Factor for the date of the current assessment and
- correcting amounts, not included elsewhere, which arise from changes to indices used for assessing previous amounts for price adjustment.

Option T: Changes in the law

Changes in the law **T1**

T1.1   A change in the law of the country in which the Site is located is a compensation event if it occurs after the Contract Date. The *Project Manager* may notify the *Contractor* of a compensation event for a change in the law and instruct him to submit quotations. If the effect of a compensation event which is a change in the law is to reduce the total Actual Cost, the Prices are reduced.

Option K: Multiple currencies (used only with Options A and B)

Multiple currencies **K1**

K1.1   The *Contractor* is paid in currencies other than the *currency of this contract* for the work listed in the Contract Data. The *exchange rates* are used to convert from the *currency of this contract* to other currencies.

K1.2   Payments to the *Contractor* in currencies other than the *currency of this contract* do not exceed the maximum amounts stated in the Contract Data. Any excess is paid in the *currency of this contract*.

Option H: Parent company guarantee

Parent company guarantee **H1**

H1.1   If a parent company owns the *Contractor*, the *Contractor* gives to the *Employer* a guarantee by the parent company of the *Contractor*'s performance in the form set out in the Works Information. If the guarantee was not given by the Contract Date, it is given to the *Employer* within four weeks of the Contract Date.

Option L: Sectional Completion

Sectional Completion **L1**

L1.1   In these *conditions of contract*, unless stated as the whole of the *works*, each reference and clause relevant to

- the *works*,
- Completion and
- Completion Date

applies, as the case may be, to either the whole of the *works* or any *section* of the *works*.

ption X2: Changes in the law

Changes in the law   X2

X2.1   A change in the law of the country in which the Site is located is a compensation event if it occurs after the Contract Date. The *Project Manager* may notify the *Contractor* of a compensation event for a change in the law and instruct him to submit quotations. If the effect of a compensation event which is a change in the law is to reduce the total Defined Cost, the Prices are reduced.

ption X3: Multiple currencies (used only with Options A and B)

Multiple currencies   X3

X3.1   The *Contractor* is paid in currencies other than the *currency of this contract* for the items or activities listed in the Contract Data. The *exchange rates* are used to convert from the *currency of this contract* to other currencies.

X3.2   Payments to the *Contractor* in currencies other than the *currency of this contract* do not exceed the maximum amounts stated in the Contract Data. Any excess is paid in the *currency of this contract*.

ption X4: Parent company guarantee

Parent company   X4
guarantee   X4.1   If a parent company owns the *Contractor*, the *Contractor* gives to the *Employer* a guarantee by the parent company of the *Contractor*'s performance in the form set out in the Works Information. If the guarantee was not given by the Contract Date, it is given to the *Employer* within four weeks of the Contract Date.

ption X5: Sectional Completion

Sectional Completion   X5

X5.1   In these *conditions of contract*, unless stated as the whole of the *works*, each reference and clause relevant to

- the *works*,
- Completion and
- Completion Date

applies, as the case may be, to either the whole of the *works* or any *section* of the *works*.

Option Q: Bonus for early Completion

Bonus for early Completion **Q1**

Q1.1 The *Contractor* is paid a bonus calculated at the rate stated in the Contr Data for each day from the earlier of

- Completion and
- the date on which the *Employer* takes over the *works*

until the Completion Date.

Option R: Delay damages

Delay damages **R1**

R1.1 The *Contractor* pays delay damages at the rate stated in the Contract Data fr the Completion Date for each day until the earlier of

- Completion and
- the date on which the *Employer* takes over the *works*.

R1.2 If the Completion Date is changed to a later date after delay damages have be paid, the *Employer* repays the overpayment of damages with interest. Interes assessed from the date of payment to the date of repayment and the date repayment is an assessment date.

Option X12: Partnering

Identified and defined terms **X12.2** (1) The Partners are those named in the Schedule of Partners. The *Client* i Partner.

(2) An Own Contract is a contract between two Partners which includes t option.

(3) The Core Group comprises the Partners listed in the Schedule of C Group Members.

(4) Partnering Information is information which specifies how the Partners wo together and is either in the documents which the Contract Data states it is in in an instruction given in accordance with the contract.

(5) A Key Performance Indicator is an aspect of performance for which a tar is stated in the Schedule of Partners.

Option X6: Bonus for early Completion

Bonus for early Completion   X6

X6.1   The *Contractor* is paid a bonus calculated at the rate stated in the Contract Data for each day from the earlier of

- Completion and
- the date on which the *Employer* takes over the *works*

until the Completion Date.

Option X7: Delay damages

Delay damages   X7

X7.1   The *Contractor* pays delay damages at the rate stated in the Contract Data from the Completion Date for each day until the earlier of

- Completion and
- the date on which the *Employer* takes over the *works*.

X7.2   If the Completion Date is changed to a later date after delay damages have been paid, the *Employer* repays the overpayment of damages with interest. Interest is assessed from the date of payment to the date of repayment and the date of repayment is an assessment date.

X7.3   If the *Employer* takes over a part of the *works* before Completion, the delay damages are reduced from the date on which the part is taken over. The *Project Manager* assesses the benefit to the *Employer* of taking over the part of the *works* as a proportion of the benefit to the *Employer* of taking over the whole of the *works* not previously taken over. The delay damages are reduced in this proportion.

> X7.3
> New subclause X7.3 provides for a situation where the *Employer* takes over part of the *works* before Completion of the whole of the *works*. The delay damages will be reduced to reflect the benefit derived by the *Employer* from that part of the *works* taken over. It is for the *Project Manager* to assess this benefit.

Option X12: Partnering

Identified and defined   X12

terms   X12.1   (1) The Partners are those named in the Schedule of Partners. The *Client* is a Partner.

(2) An Own Contract is a contract between two Partners which includes this Option.

(3) The Core Group comprises the Partners listed in the Schedule of Core Group Members.

(4) Partnering Information is information which specifies how the Partners work together and is either in the documents which the Contract Data states it is in or in an instruction given in accordance with this contract.

(5) A Key Performance Indicator is an aspect of performance for which a target is stated in the Schedule of Partners.

Actions X12.1 (1) Each Partner works with the other Partners to achieve the *Client's* objective stated in the Contract Data and the objectives of every other Partner stated in the Schedule of Partners.

(2) Each Partner nominates a representative to act for it in dealings with other Partners.

(3) The Core Group acts and takes decisions on behalf of the Partners on those matters stated in the Partnering Information.

(4) The Partners select the members of the Core Group. The Core Group decides how they will work and decides the dates when each member joins and leaves the Core Group. The *Client's* representative leads the Core Group unless stated otherwise in the Partnering Information.

(5) The Core Group keeps the Schedule of Core Group Members and the Schedule of Partners up to date and issues copies of them to the Partners each time either is revised.

(6) This option does not create a legal partnership between Partners who are not one of the Parties in this contract.

Working together X12.3 (1) The Partners work together as stated in the Partnering Information and in a spirit of mutual trust and cooperation.

(2) A Partner may ask another Partner to provide information that he needs to carry out the work in his Own Contract and the other Partner provides it.

(3) Each Partner gives an early warning to the other Partners when he becomes aware of any matter that could affect the achievement of another Partner's objectives stated in the Schedule of Partners.

(4) The Partners use common information systems as set out in the Partnering Information.

(5) A Partner implements a decision of the Core Group by issuing instructions in accordance with its Own Contracts.

(6) The Core Group may give an instruction to the Partners to change the Partnering Information. Each such change to the Partnering Information is a compensation event which may lead to reduced Prices.

(7) The Core Group prepares and maintains a timetable showing the proposed timing of the contributions of the Partners. The Core Group issues a copy of the timetable to the Partners each time it is revised. A Partner incorporates information in the timetable into its Own Contract programme.

(8) A Partner gives advice, information and opinion to the Core Group and to other Partners when asked to do so by the Core Group. This advice, information and opinion relates to work that the other Partner is carrying out under its Own Contract and is given fully, openly and objectively. The Partners show contingency and risk allowances in information about costs, prices and timing for future work.

(9) A Partner notifies the Core Group before subcontracting any work. A Partner is responsible under its Own Contract for the actions and inactions of its subcontractor.

Incentives X12.4 (1) A Partner is paid the amount stated in the Schedule of Partners if the target stated for a Key Performance Indicator is improved upon or achieved. Payment of the amount is due when the target has been improved upon or achieved and is made as part of the amount due in the Partner's Own Contract.

(2) The *Client* may add a Key Performance Indicator or associated payment but may not delete or reduce a payment stated in the Schedule of Partners.

Actions   X12.2   (1) Each Partner works with the other Partners to achieve the *Client's objective* stated in the Contract Data and the objectives of every other Partner stated in the Schedule of Partners.

(2) Each Partner nominates a representative to act for it in dealings with other Partners.

(3) The Core Group acts and takes decisions on behalf of the Partners on those matters stated in the Partnering Information.

(4) The Partners select the members of the Core Group. The Core Group decides how they will work and decides the dates when each member joins and leaves the Core Group. The *Client's* representative leads the Core Group unless stated otherwise in the Partnering Information.

(5) The Core Group keeps the Schedule of Core Group Members and the Schedule of Partners up to date and issues copies of them to the Partners each time either is revised.

(6) This Option does not create a legal partnership between Partners who are not one of the Parties in this contract.

Working together   X12.3   (1) The Partners work together as stated in the Partnering Information and in a spirit of mutual trust and co-operation.

(2) A Partner may ask another Partner to provide information which he needs to carry out the work in his Own Contract and the other Partner provides it.

(3) Each Partner gives an early warning to the other Partners when he becomes aware of any matter that could affect the achievement of another Partner's objectives stated in the Schedule of Partners.

(4) The Partners use common information systems as set out in the Partnering Information.

(5) A Partner implements a decision of the Core Group by issuing instructions in accordance with its Own Contracts.

(6) The Core Group may give an instruction to the Partners to change the Partnering Information. Each such change to the Partnering Information is a compensation event which may lead to reduced Prices.

(7) The Core Group prepares and maintains a timetable showing the proposed timing of the contributions of the Partners. The Core Group issues a copy of the timetable to the Partners each time it is revised. The *Contractor* changes his programme if it is necessary to do so in order to comply with the revised timetable. Each such change is a compensation event which may lead to reduced Prices.

A new compensation event has been introduced for when the *Contractor* has to change his programme in order to comply with the revised timetable.

(8) A Partner gives advice, information and opinion to the Core Group and to other Partners when asked to do so by the Core Group. This advice, information and opinion relates to work that another Partner is to carry out under its Own Contract and is given fully, openly and objectively. The Partners show contingency and risk allowances in information about costs, prices and timing for future work.

(9) A Partner notifies the Core Group before subcontracting any work.

Incentives   X12.4   (1) A Partner is paid the amount stated in the Schedule of Partners if the target stated for a Key Performance Indicator is improved upon or achieved. Payment of the amount is due when the target has been improved upon or achieved and is made as part of the amount due in the Partner's Own Contract.

(2) The *Client* may add a Key Performance Indicator and associated payment to the Schedule of Partners but may not delete or reduce a payment stated in the Schedule of Partners.

Option G: Performance bond

Performance bond **G1**

G1.1 The *Contractor* gives the *Employer* a performance bond, provided by a bank or insurer which the *Project Manager* has accepted, for the amount stated in the Contract Data and in the form set out in the Works Information. A reason for not accepting the bank or insurer is that its commercial position is not strong enough to carry the bond. If the bond was not given by the Contract Date, it is given to the *Employer* within four weeks of the Contract Date.

Option J: Advanced payment to the *Contractor*

Advanced payment **J1**

J1.1 The *Employer* makes an advanced payment to the *Contractor* of the amount stated in the Contract Data.

J1.2 The advanced payment is made either within four weeks of the Contract Date or, if an advanced payment bond is required, within four weeks of the later of

- the Contract Date and

- the date when the *Employer* receives the advanced payment bond.

The advanced payment bond is issued by a bank or insurer which the *Project Manager* has accepted. A reason for not accepting the proposed bank or insurer is that its commercial position is not strong enough to carry the bond. The bond is for the amount of the advanced payment and in the form set out in the Works Information. Delay in making the advanced payment is a compensation event.

J1.3 The advanced payment is repaid to the *Employer* by the *Contractor* in instalments of the amount stated in the Contract Data. An instalment is included in each amount due assessed after the period stated in the Contract Data has passed until the advanced payment has been repaid.

Option M: Limitation of the *Contractor*'s liability for his design to reasonable skill and care

The *Contractor*'s design **M1**

M1.1 The *Contractor* is not liable for Defects in the *works* due to his design so far as he proves that he used reasonable skill and care to ensure that it complied with the Works Information.

ption X13: Performance bond

Performance bond X13

X13.1 The *Contractor* gives the *Employer* a performance bond, provided by a bank or insurer which the *Project Manager* has accepted, for the amount stated in the Contract Data and in the form set out in the Works Information. A reason for not accepting the bank or insurer is that its commercial position is not strong enough to carry the bond. If the bond was not given by the Contract Date, it is given to the *Employer* within four weeks of the Contract Date.

ption X14: Advanced payment to the *Contractor*

Advanced payment X14

X14.1 The *Employer* makes an advanced payment to the *Contractor* of the amount stated in the Contract Data.

X14.2 The advanced payment is made either within four weeks of the Contract Date or, if an advanced payment bond is required, within four weeks of the later of

- the Contract Date and
- the date when the *Employer* receives the advanced payment bond.

The advanced payment bond is issued by a bank or insurer which the *Project Manager* has accepted. A reason for not accepting the proposed bank or insurer is that its commercial position is not strong enough to carry the bond. The bond is for the amount of the advanced payment which the *Contractor* has not repaid and is in the form set out in the Works Information. Delay in making the advanced payment is a compensation event.

X14.3 The advanced payment is repaid to the *Employer* by the *Contractor* in instalments of the amount stated in the Contract Data. An instalment is included in each amount due assessed after the period stated in the Contract Data has passed until the advanced payment has been repaid.

ption X15: Limitation of the *Contractor*'s liability for his design to reasonable skill and care

The *Contractor*'s design X15

X15.1 The *Contractor* is not liable for Defects in the *works* due to his design so far as he proves that he used reasonable skill and care to ensure that his design complied with the Works Information.

In X15.1 'his design' replaces 'it' to clarify the basis of the test the *Contractor* may have to conduct.

X15.2 If the *Contractor* corrects a Defect for which he is not liable under this contract it is a compensation event.

This new subclause adds to the list of possible compensation events.

Option P: Retention (used only with Options A, B, C, D and E)

Retention **P1**

P1.1   After the Price for Work Done to Date has reached the *retention free amount*, an amount is retained in each amount due assessed. Until the earlier of

- Completion of the whole of the *works* and
- the date on which the *Employer* takes over the whole of the *works*

the amount retained is the *retention percentage* applied to the excess of the Price for Work Done to Date above the *retention free amount*.

P1.2   The amount retained is halved

- in the assessment made at Completion of the whole of the *works* or
- in the next assessment after the *Employer* has taken over the whole of the *works* if this is before Completion of the whole of the *works*.

The amount retained remains at this amount until the Defects Certificate is issued. No amount is retained in the assessments made after the Defects Certificate has been issued.

Option S: Low performance damages

Low performance damages **S1**

S1.1   If a Defect included in the Defects Certificate shows low performance with respect to a performance level stated in the Contract Data, the *Contractor* pays the amount of low performance damages stated in the Contract Data.

21.5   The *Contractor*'s liability to the *Employer* for Defects due to his design that are not listed on the Defects Certificate is limited to the amount stated in the Contract Data in addition to any damages stated in this contract for delay or low performance.

Option X16: Retention (not used with Option F)

Retention   **X16**

X16.1   After the Price for Work Done to Date has reached the *retention free amount*, an amount is retained in each amount due. Until the earlier of

- Completion of the whole of the *works* and
- the date on which the *Employer* takes over the whole of the *works*

the amount retained is the *retention percentage* applied to the excess of the Price for Work Done to Date above the *retention free amount*.

X16.2   The amount retained is halved

- in the assessment made at Completion of the whole of the *works* or
- in the next assessment after the *Employer* has taken over the whole of the *works* if this is before Completion of the whole of the *works*.

The amount retained remains at this amount until the Defects Certificate is issued. No amount is retained in the assessments made after the Defects Certificate has been issued.

Option X17: Low performance damages

Low performance   **X17**
damages

X17.1   If a Defect included in the Defects Certificate shows low performance with respect to a performance level stated in the Contract Data, the *Contractor* pays the amount of low performance damages stated in the Contract Data.

Option X18: Limitation of liability

Limitation of liability   **X18**

X18.1   The *Contractor*'s liability to the *Employer* for the *Employer*'s indirect or consequential loss is limited to the amount stated in the Contract Data.

X18.2   For any one event, the liability of the *Contractor* to the *Employer* for loss of or damage to the *Employer*'s property is limited to the amount stated in the Contract Data.

X18.3   The *Contractor*'s liability to the *Employer* for Defects due to his design which are not listed on the Defects Certificate is limited to the amount stated in the Contract Data.

X18.4   The *Contractor*'s total liability to the *Employer* for all matters arising under or in connection with this contract, other than the excluded matters, is limited to the amount stated in the Contract Data and applies in contract, tort or delict and otherwise to the extent allowed under the *law of the contract.*

The excluded matters are amounts payable by the *Contractor* as stated in this contract for

- loss of or damage to the *Employer*'s property,
- delay damages if Option X7 applies,
- low performance damages if Option X17 applies and
- *Contractor*'s share if Option C or Option D applies.

Option U: The Construction (Design and Management) Regulations 1994 (to be used for contracts in UK)

The CDM Regulations 1994   U1.1   A delay to the work or additional or changed work caused by application of The Construction (Design and Management) Regulations 1994 is a compensation event if an experienced contractor could not reasonably be expected to have foreseen it.

Option V: Trust Fund

Defined terms   **V1**

V1.1   (1) The Trust Fund is a fund held and administered by the *Trustees*.

(2) The Trust Deed is a deed between the *Employer* and the *Trustees* which contains the provisions for administering the Trust Fund. Terms defined in this contract have the same meaning in the Trust Deed.

(3) The Initial Value of the Trust Fund is an amount which is the total of the Prices at the Contract Date multiplied by 1.5 and divided by the number of months in the period between the Contract Date and the Completion Date.

(4) Insolvency of an individual occurs when that individual has

- presented his petition for bankruptcy,
- had a bankruptcy order made against him,
- had a receiver appointed over his assets or
- made an arrangement with his creditors.

(5) Insolvency of a company occurs when it has

- had a winding-up order made against it,
- had a provisional liquidator appointed to it,
- passed a resolution for winding-up (other than in order to amalgamate or reconstruct),
- had an administration order made against it,
- had a receiver, receiver and manager, or administrative receiver appointed over the whole or a substantial part of its undertaking or assets or
- made an arrangement with its creditors.

(6) The Beneficiaries are the *Contractor* and

- Subcontractors,
- suppliers of the *Contractor*,
- subcontractors of whatever tier of a Subcontractor and
- suppliers of whatever tier of a Subcontractor or of his subcontractors

who are employed to Provide the Works.

(7) A Trust Payment is a payment made by the *Trustees* out of the Trust Fund.

ECC2 Options U and V have been deleted in ECC3. Option U set an unclear split of risk between the Parties and the trust fund provisions in Option V were unpopular in the industry. Should a trust fund be required in ECC3 then the ECC2 Option V provisions could offer a basis for drafting suitable Option Z *additional condition of contract*.

X18.5   The *Contractor* is not liable to the *Employer* for a matter unless it is notified to the *Contractor* before the *end of liability date*.

An entirely new clause has been added due to demands from the contracting industry and their insurers to cap some of the potential losses and set time limits in which liability may arise following certain events. The clause itself is optional, as are parts of it. Parties may therefore agree to include none, part or all of the provisions through appropriate completion of the Contract Data.

The matters of limitation include the *Employer*'s indirect or consequential loss, loss of or damage to the *Employer*'s property and *Employer*'s loss due to Defects arising from the *Contractor*'s design (note ECC2 subclause 21.5 has been moved to X18.3). These are expressed as individual liabilities in the Contract Data but there is also a cap of total liability to the *Employer* applying in contract, tort or delict, except for listed excluded matters which sit outside this cap.

There is a time constraint on application of this clause in X18.5. The *Employer* must notify such matters before the *end of liability date*, this also being stated in the Contract Data. This can usefully add a cut-off date of liability in jurisdictions without such provisions in law, or could reduce such periods where the law does provide this.

Option X20: Key Performance Indicators (not used with Option X12)

Incentives   X20.1   A Key Performance Indicator is an aspect of performance by the *Contractor* for which a target is stated in the Incentive Schedule. The Incentive Schedule is the *incentive schedule* unless later changed in accordance with this contract.

X20.2   From the *starting date* until the Defects Certificate has been issued, the *Contractor* reports to the *Project Manager* his performance against each of the Key Performance Indicators. Reports are provided at the intervals stated in the Contract Data and include the forecast final measurement against each indicator.

X20.3   If the *Contractor*'s forecast final measurement against a Key Performance Indicator will not achieve the target stated in the Incentive Schedule, he submits to the *Project Manager* his proposals for improving performance.

X20.4   The *Contractor* is paid the amount stated in the Incentive Schedule if the target stated for a Key Performance Indicator is improved upon or achieved. Payment of the amount is due when the target has been improved upon or achieved.

Trust Fund V2

V2.1 The *Employer* establishes the Trust Fund within one week of the Contract Date

V2.2 The Trust Fund is established

- by the *Employer* making a payment to the *Trustees* equal to the Initi Value or

- by the *Employer* providing the *Trustees* with a guarantee of the Initi Value, payable to the *Trustees* on their first written demand, given by bank or other financial institution acceptable to the *Trustees* or,

- if the *Employer* is a Government department or other public authority i the United Kingdom, by the *Employer* entering into irrevocable unde takings with the *Contractor* and the *Trustees* to pay the *Trustees* prompt on demand such amounts as they request for

  - Trust Payments and

  - their fees and expenses for administering the Trust Fund.

V2.3 The *Contractor* informs his suppliers and his Subcontractors of the terms of th Trust Deed and of the appointment of the *Trustees*. He arranges tha Subcontractors ensure that their suppliers and subcontractors, of whatever tie are also informed.

Trust Deed V3

V3.1 The Trust Fund is administered by the *Trustees* in accordance with the Tru Deed. The Trust Deed includes the following provisions.

(1) If a Beneficiary satisfies the *Trustees*

- that he has not received all or part of a payment properly due to hi under his contract relating to the *works* which was unpaid at the time the Insolvency and

- that the reason for the failure to pay is the Insolvency of the party whic should have made the payment

the *Trustees* may at their discretion make a Trust Payment to the Beneficiary an amount not exceeding the value of the payment which he has not received.

(2) If a Beneficiary subsequently receives a payment from another party, i respect of which a Trust Payment has been made, the Beneficiary passes on tha payment to the *Trustees* (up to the value of the Trust Payment). Before making Trust Payment the *Trustees* may require from a Beneficiary either an assignmer of rights or an undertaking with respect to that payment in a form acceptable t them.

(3) The *Trustees* have discretion to decide the amount and timing of every Tru Payment. They may make a Trust Payment on account or withhold a Trust Pa ment until they have assessed the total amount of debts owing to a Beneficiar arising out of an Insolvency. They may take into account any claims (includin claims by way of set-off) which the party suffering from Insolvency may hav against the Beneficiary as well as the likely ability of the liquidator or othe administrator of the insolvent party to meet the claims of unsecured credito from funds in his hands.

(4) If the Trust Fund was established by a payment, the *Employer* maintains th Trust Fund at the Initial Value. If the Trust Fund was established by guarantor, the *Employer* ensures that the guarantor maintains the Trust Fund the Initial Value. The *Trustees* notify the *Employer* or the guarantor within or week of making a Trust Payment and the *Employer* or the guarantor restor the Trust Fund to the Initial Value within two weeks of the notification.

X20.5   The *Employer* may add a Key Performance Indicator and associated payment to the Incentive Schedule but may not delete or reduce a payment stated in the Incentive Schedule.

A further new secondary Option reflecting best practice procurement. Key Performance Indicators (KPIs) were available in ECC2 Option X12 but only for use in multi-party partnering arrangements. Option X20 is provided for use in a bi-party arrangement. The focus is the *Contractor* achieving the targets stated in the Incentive Schedule and in turn receiving the associated payment accordingly. Regular reports are provided to the *Project Manager* together with proposals for improving performance where the forecast is not to achieve the stated target. There are no guide KPIs provided, this is entirely down to parties to agree appropriate Indicators to reflect project objectives or any other targets they desire. KPIs are a tool to assist the process of continuous improvement and have been drafted in this manner and not as a penalty deduction if the target is not ultimately met.

(5) After the *Trustees* have made all Trust Payments, any amount in the Trust Fund (including any accrued interest) is paid by the *Trustees* to the *Employer*. If a guarantee has been provided, it is returned to the guarantor. The *Trustees* do not pay claims from Beneficiaries which they receive after the Defects Certificate has been issued.

(6) The *Employer* pays the *Trustees* their fees and expenses for administering the Trust Fund.

(7) The *Trustees* may engage professional consultants to help them with the administration of the Trust Fund and may make Trust Payments for their fees and expenses.

(8) The *Trustees* hold the Trust Fund on an interest-bearing bank account.

Addendum Y(UK)2 Contract (Ref NEC/ECC/Y(UK)2/April 1998) to take into account The Housing Grants, Construction and Regeneration Act 1996 (Part II).

Option Y(UK)2: The Housing Grants, Construction and Regeneration Act 1996

Y2.1 In this Option

- the Act means The Housing Grants, Construction and Regeneration Act 1996 and

- periods of time stated in days are reckoned in accordance with Section 116 of the Act.

Y2.2 **Clause 51 is amended as follows:**

Clause 51.1 the first sentence is deleted and replaced with the following sentence

'The *Project Manager* certifies a payment on or before the date on which a payment becomes due.'

Clause 51.2 the first sentence is deleted and replaced with the following sentence:-

'Each certified payment is made on or before the final date for payment.'

Y2.3 **The following clauses are added**

Dates for Payment 56
56.1 For the purpose of Sections 109 and 110 of the Act,

- the *Project Manager*'s certificate is the notice of payment from the *Employer* to the *Contractor* specifying the amount (if any) of the payment made or proposed to be made, and the basis on which that amount was calculated,

- the date on which a payment becomes due is seven days after the assessment date and

PTION Y

ption Y(UK)2: The Housing Grants, Construction and Regeneration Act 1996

Definitions   **Y(UK)2**

Y2.1   (1) The Act is The Housing Grants, Construction and Regeneration Act 1996.

(2) A period of time stated in days is a period calculated in accordance with Section 116 of the Act.

Dates for payment   Y2.2   The date on which a payment becomes due is seven days after the assessment date.

The final date for payment is fourteen days or a different period for payment if stated in the Contract Data after the date on which payment becomes due.

The *Project Manager*'s certificate is the notice of payment from the *Employer* to the *Contractor* specifying the amount of the payment made or proposed to be made and stating how the amount was calculated.

- the final date for payment is
  - twenty one days or
  - if a different period for payment is stated in the Contract Data, period stated

after the date on which the payment becomes due.

56.2 If the *Employer* intends to withhold payment after the final date for payment c sum due under this contract, he notifies the *Contractor* not later than seven d: (the prescribed period) before the final date for payment by specifying

- the amount proposed to be withheld and the ground for withholding p ment or
- if there is more than one ground, each ground and the amount attributa to it.

Y2.4 **The following is added to clause 60**

60.7 Suspension of performance is a compensation event if the *Contractor* exerci his right to suspend performance under the Act.

Addendum Y(UK)3 (Ref NEC/Y(UK)3/April 2000) to take into account The Contracts (Rights of Third Parties) Act 1999.

Option Y(UK)3: The Contracts (Rights of Third Parties) Act 1999

Y3.1 For the purposes of the Contracts (Rights of Third Parties) Act 1999, nothing this contract confers or purports to confer on a third party any benefit or a right to enforce a term of this contract.

Option Z: Additional conditions of contract

Additional conditions of **Z1**
contract Z1.1 The additional conditions of contract stated in the Contract Data are part this contract.

Notice of intention to withhold payment   Y2.3   If either Party intends to withhold payment of an amount due under this contract, he notifies the other Party not later than seven days (the prescribed period) before the final date for payment by stating the amount proposed to be withheld and the reason for withholding payment. If there is more than one reason, the amount for each reason is stated.

A Party does not withhold payment of an amount due under this contract unless he has notified his intention to withhold payment as required by this contract.

Suspension of performance   Y2.4   If the *Contractor* exercises his right under the Act to suspend performance, it is a compensation event.

These provisions deal only with the aspects of the Housing, Grants, Construction and Regeneration Act 1996 not dealt with in the dispute resolution Options W1 and W2. Y(UK)2 under ECC2 covered both disputes and payment provisions under this Act. These supplementary clauses are only of relevance in the UK and cover due date for payment and final date for payment provisions together with other features that must be included in a contract covered by this Act.

tion Y(UK)3: The Contracts (Rights of Third Parties) Act 1999

Third party rights   Y(UK)3

Y3.1   A person or organisation who is not one of the Parties may enforce a term of this contract under the Contracts (Rights of Third Parties) Act 1999 only if the term and the person or organisation are stated in the Contract Data.

In ECC2 the inclusion of this Option was intended to avoid giving third parties any rights to which they are not a party to under a contract. The ECC3 use of this Option is the opposite way round, with the provision being to expressly state in the Contract Data the term of the contract and the person or organisation that the third party may enforce.

tion Z: *Additional conditions of contract*

Additional conditions of contract   **Z1**

Z1.1   The *additional conditions of contract* stated in the Contract Data are part of this contract.

SCHEDULE OF COST COMPONENTS

This schedule is not part of the *conditions of contract* when Option F is us
When Option C, D or E is used, in this schedule the *Contractor* means
Contractor and not his Subcontractors. Amounts are included only in one c
component.

People 1   The following components of the cost of

- people who are directly employed by the *Contractor* and whose norr place of working is within the Working Areas,
- people who are directly employed by the *Contractor* and whose norr place of working is not within the Working Areas but who are working the Working Areas for a period of not less than one week and
- people who are not directly employed by the *Contractor* but are paid the *Contractor* according to the time worked while they are within Working Areas.

11   Wages and salaries.

12   Payments to people for

    (a) bonuses and incentives
    (b) overtime
    (c) working in special circumstances
    (d) special allowances
    (e) absence due to sickness and holidays
    (f) severance related to work on this contract.

13   Payments made in relation to people for

    (a) travelling to and from the Working Areas
    (b) subsistence and lodging
    (c) relocation
    (d) medical examinations
    (e) passports and visas
    (f) travel insurance
    (g) items (a) to (f) for a spouse or dependents
    (h) protective clothing
    (j) meeting the requirements of the law
    (k) superannuation and life assurance
    (l) death benefit
    (m) occupational accident benefits
    (n) medical aid.

SCHEDULE OF COST COMPONENTS

This schedule is part of the *conditions of contract* only when Option C, D or E is used. In this schedule the *Contractor* means the *Contractor* and not his Subcontractors. An amount is included only in one cost component and only if it is incurred in order to Provide the Works.

The additional text at the end of the second sentence emphasises that in the administration of cost components amounts must be incurred in order to 'Provide the Works' and the *Employer* should not be expected to pay, for example, for idle plant stood on the Site for the *Contractor*'s convenience. This Schedule does not feature in the administration of main Option A or B contracts.

People 1 The following components of the cost of

- people who are directly employed by the *Contractor* and whose normal place of working is within the Working Areas and
- people who are directly employed by the *Contractor* and whose normal place of working is not within the Working Areas but who are working in the Working Areas.

11 Wages, salaries and amounts paid by the *Contractor* for people paid according to the time worked while they are within the Working Areas.

12 Payments to people for

(a) bonuses and incentives
(b) overtime
(c) working in special circumstances
(d) special allowances
(e) absence due to sickness and holidays
(f) severance related to work on this contract.

13 Payments made in relation to people for

(a) travel
(b) subsistence and lodging
(c) relocation
(d) medical examinations
(e) passports and visas
(f) travel insurance
(g) items (a) to (f) for dependants
(h) protective clothing
(i) meeting the requirements of the law
(j) pensions and life assurance
(k) death benefit
(l) occupational accident benefits
(m) medical aid
(n) a vehicle
(o) safety training.

14 The following components of the cost of people who are not directly employed by the *Contractor* but are paid for by him according to the time worked while they are within the Working Areas.

Amounts paid by the *Contractor*.

The ECC2 5-day qualifying rule for payment of people who are directly employed by the *Contractor* and whose normal place of working is not within the Working Areas has been deleted. Occasional or regular part-time *Contractor*'s people are now a recoverable cost under ECC3. This should lead to easier administration and reduce the Fee.

Equipment 2 The following components of the cost of Equipment which is used within th Working Areas (excluding Equipment cost covered by the percentage fc Working Areas overheads).

21 Payments for the hire of Equipment not owned by the *Contractor*, by th *Contractor*'s parent company or by another part of a group with the sam parent company.

22 An amount for depreciation and maintenance of Equipment which is

    (a) owned by the *Contractor*
    (b) purchased by the *Contractor* under a hire purchase or lease agreement o
    (c) hired by the *Contractor* from the *Contractor*'s parent company c another part of a group with the same parent company.

The depreciation and maintenance charge is the actual purchase price of th item of Equipment (or first cost if the *Contractor* assembled, fabricated or othe wise produced the item of Equipment) divided by its average working lit remaining at the time of purchase, assembly or fabrication (expressed in weeks First cost is limited to the cost of manufacture when the item of Equipmer came into being.

The amount for depreciation and maintenance is calculated by multiplying th depreciation and maintenance charge by the time required and then increasin the product by the appropriate percentage for Equipment depreciation an maintenance stated in the Contract Data.

The time required is the number of weeks and part weeks. A part week measured in half days and is expressed as one twelfth of a week. A part half da is taken as a half day. For idle or standby time the first half day is deducted.

23 The purchase price of Equipment which is consumed.

24 Unless included in the hire rates or the depreciation and maintenance charge payments for

    (a) transporting Equipment to and from the Working Areas
    (b) erecting and dismantling Equipment
    (c) upgrading or modification needed for a compensation event.

25 Unless included in the hire rates or the depreciation and maintenance charge the cost of operatives is included in the cost of people.

This new provision covers the likes of resources commonly referred to as "labour only subcontractors" and its effect is quite different to the equivalent under ECC2. ECC2 provisions meant pairing back the invoiced cost of such labour to comply with those people costs recoverable under 11, 12 and 13. Such information was not readily forthcoming in practice. This approach takes the amount paid by the *Contractor* as the basis of cost subject of course to the subclause 52.1 open market prices criteria. Again, this should ease administration and reduce the Fee.

Equipment 2 The following components of the cost of Equipment which is used within the Working Areas (including the cost of accommodation but excluding Equipment cost covered by the percentage for Working Areas overheads).

21 Payments for the hire or rent of Equipment not owned by

- the *Contractor*,
- his parent company or
- by a company with the same parent company

at the hire or rental rate multiplied by the time for which the Equipment is required.

22 Payments for Equipment which is not listed in the Contract Data but is

- owned by the *Contractor*,
- purchased by the *Contractor* under a hire purchase or lease agreement or
- hired by the *Contractor* from the *Contractor*'s parent company or from a company with the same parent company

at open market rates, multiplied by the time for which the Equipment is required.

23 Payments for Equipment purchased for work included in this contract listed with a time-related on cost charge, in the Contract Data, of

- the change in value over the period for which the Equipment is required and
- the time-related on cost charge stated in the Contract Data for the period for which the Equipment is required.

The change in value is the difference between the purchase price and either the sale price or the open market sale price at the end of the period for which the Equipment is required. Interim payments of the change in value are made at each assessment date. A final payment is made in the next assessment after the change in value has been determined.

If the *Project Manager* agrees, an additional item of Equipment may be assessed as if it had been listed in the Contract Data.

24 Payments for special Equipment listed in the Contract Data. These amounts are the rates stated in the Contract Data multiplied by the time for which the Equipment is required.

If the *Project Manager* agrees, an additional item of special Equipment may be assessed as if it had been listed in the Contract Data.

25 Payments for the purchase price of Equipment which is consumed.

26 Unless included in the hire or rental rates, payments for

- transporting Equipment to and from the Working Areas other than for repair and maintenance,
- erecting and dismantling Equipment and
- constructing, fabricating or modifying Equipment as a result of a compensation event.

27 Payments for purchase of materials used to construct or fabricate Equipment.

28 Unless included in the hire rates, the cost of operatives is included in the cost of people.

Plant and Materials 3   The following components of the cost of Plant and Materials.

31   Payments for

(a) purchasing Plant and Materials
(b) delivery to and removal from the Working Areas
(c) providing and removing packaging
(d) samples and tests.

32   Cost is credited with payments received for disposal of Plant and Materials.

Charges 4   The following components of the cost of charges paid by the *Contractor*.

41   Payments to utilities for provision and use in the Working Areas of

(a) water
(b) gas
(c) electricity
(d) other services.

42   Payments to public authorities, utilities and other properly constituted authoritie of charges which they are authorised to make in respect of the *works*.

43   Payments for

(a) financing charges (excluding charges compensated for by interest paid i accordance with this contract)
(b) buying or leasing land
(c) compensation for loss of crops or buildings
(d) royalties
(e) inspection certificates
(f) rent of premises in the Working Areas
(g) charges for access to the Working Areas
(h) facilities for visits to the Working Areas by Others
(i) specialist services.

The ECC2 Equipment provisions have been substantially amended in ECC3, leading to a more practical application. ECC2 necessitated a quite complex calculation of depreciation and maintenance for *Contractor*-owned, bought or internally hired plant. This is now replaced in ECC3 with open market rates or rates stated in the Contract Data as applicable for the equivalent plant.

A further change in ECC3 is to bring in the cost of accommodation as a recoverable cost under the heading Equipment. The ECC2 items 44(a) to (f) have been deleted in this change. This should provide a more equitable basis of recovering these costs rather than by linking the Working Areas overheads percentage to the cost of people. This will also result in this percentage being substantially reduced from the level it was in ECC2 and reducing the uncertainty surrounding it.

Plant and Materials **3**   The following components of the cost of Plant and Materials.

          31   Payments for

- purchasing Plant and Materials,
- delivery to and removal from the Working Areas,
- providing and removing packaging and
- samples and tests.

          32   Cost is credited with payments received for disposal of Plant and Materials unless the cost is disallowed.

Charges **4**   The following components of the cost of charges paid by the *Contractor.*

          41   Payments for provision and use in the Working Areas of

- water,
- gas and
- electricity.

          42   Payments to public authorities and other properly constituted authorities of charges which they are authorised to make in respect of the *works*.

          43   Payments for

    (a) cancellation charges arising from a compensation event
    (b) buying or leasing land
    (c) compensation for loss of crops or buildings
    (d) royalties
    (e) inspection certificates
    (f) charges for access to the Working Areas
    (g) facilities for visits to the Working Areas by Others
    (h) specialist services
    (i) consumables and equipment provided by the *Contractor* for the *Project Manager*'s and *Supervisor*'s offices.

The ECC2 provisions for items 43 have been amended. Item 43(a) 'financing charges' has been deleted as a recoverable component of cost and is therefore treated as included in the Fee. This is replaced with 'cancellation charges arising from a compensation event', picking up for example any such charges levied by a change to the Works Information that necessitates cancelling a previous order by the *Contractor*. New item 43(i) deals with the consumables and equipment for *Project Manager*'s and *Supervisor*'s offices.

44   A charge for overhead costs incurred within the Working Areas calculated by applying the percentage for Working Areas overheads stated in the Contract Data to the total of people items 11, 12 and 13. The charge includes provision and use of accommodation, equipment, supplies and services for

(a) offices and drawing offices
(b) laboratories
(c) workshops
(d) stores and compounds
(e) labour camps
(f) cabins
(g) catering
(h) medical facilities and first aid
(j) recreation
(k) sanitation
(l) security
(m) copying
(n) telephone, telex, fax, radio and CCTV
(o) surveying and setting out
(p) computing
(q) hand tools and hand held powered tools.

Manufacture and fabrication 5   The following components of the cost of manufacture or fabrication of Plant and Materials which are

- wholly or partly designed specifically for the *works* and

- manufactured or fabricated outside the Working Areas.

51   The total of the hours worked by employees multiplied by the hourly rates stated in the Contract Data for the categories of employees listed.

52   An amount for overheads calculated by multiplying this total by the percentage for manufacturing and fabrication overheads stated in the Contract Data.

Design 6   The following components of the cost of design of the *works* and Equipment done outside the Working Areas.

61   The total of the hours worked by employees multiplied by the hourly rates stated in the Contract Data for the categories of employees listed.

62   An amount for overheads calculated by multiplying this total by the percentage for design overheads stated in the Contract Data.

63   The cost of travel to and from the Working Areas for the categories of employees listed in the Contract Data.

Insurance 7   The following are deducted from cost:

- the cost of events for which this contract requires the *Contractor* to insure and

- other costs paid to the *Contractor* by insurers.

44   A charge for overhead costs incurred within the Working Areas calculated by applying the percentage for Working Areas overheads stated in the Contract Data to the total of people items 11, 12, 13 and 14. The charge includes provision and use of equipment, supplies and services, but excludes accommodation, for

(a) catering
(b) medical facilities and first aid
(c) recreation
(d) sanitation
(e) security
(f) copying
(g) telephone, telex, fax, radio and CCTV
(h) surveying and setting out
(i) computing
(j) hand tools not powered by compressed air.

A change to item 44(j) results in any hand tools powered by compressed air being items of Equipment and so included with those that are recovered through application of the Working Areas overheads percentage.

Manufacture and fabrication   **5**   The following components of the cost of manufacture and fabrication of Plant and Materials which are

• wholly or partly designed specifically for the *works* and
• manufactured or fabricated outside the Working Areas.

51   The total of the hours worked by employees multiplied by the hourly rates stated in the Contract Data for the categories of employees listed.

52   An amount for overheads calculated by multiplying this total by the percentage for manufacturing and fabrication overheads stated in the Contract Data.

Design   **6**   The following components of the cost of design of the *works* and Equipment done outside the Working Areas.

61   The total of the hours worked by employees multiplied by the hourly rates stated in the Contract Data for the categories of employees listed.

62   An amount for overheads calculated by multiplying this total by the percentage for design overheads stated in the Contract Data.

63   The cost of travel to and from the Working Areas for the categories of design employees listed in the Contract Data.

Insurance   **7**   The following are deducted from cost

• the cost of events for which this contract requires the *Contractor* to insure and
• other costs paid to the *Contractor* by insurers.

SHORTER SCHEDULE OF COST COMPONENTS

This schedule is not part of the *conditions of contract* when Option F is used When Option C, D or E is used, in this schedule the *Contractor* means th *Contractor* and not his Subcontractors. Amounts are included only in one co component.

People 1 The following components of the cost of

- people who are directly employed by the *Contractor* and whose norma place of working is within the Working Areas and

- people who are directly employed by the *Contractor* and whose norma place of working is not within the Working Areas but who are working i the Working Areas for a period of not less than one week and

- people who are not directly employed by the *Contractor* but are paid b the *Contractor* according to the time worked while they are within th Working Areas.

11 Wages and salaries.

12 Payments to people for

    (a) bonuses and incentives
    (b) overtime
    (c) working in special circumstances
    (d) special allowances.

13 Payments made in relation to people for

    (a) travelling to and from the Working Areas
    (b) subsistence and lodging.

Equipment 2 The following components of the cost of Equipment used within the Workin Areas (excluding Equipment cost covered by the percentage for people overheads)

21 Amounts for Equipment which is in the published list stated in the Contrac Data. These amounts are calculated by applying the percentage adjustment fc listed Equipment stated in the Contract Data to the rates in the published list an by multiplying the resulting rate by the time for which the Equipment is required

22 Amounts for Equipment listed in the Contract Data which is not in th published list stated in the Contract Data. These amounts are the rates stated i the Contract Data multiplied by the time for which the Equipment is required.

23 The time required is expressed in hours, days, weeks or months consistently wit the list of items of Equipment in the Contract Data or with the published li stated in the Contract Data. For idle and standby time, the following times ar deducted:

- the first two hours for items paid at an hourly rate,

- the first half day for items paid at a daily rate,

- the first third of a week for items paid at a weekly rate and

- the first quarter of a month for items paid at a monthly rate.

24 Unless the item is in the published list and the rate includes the cost componen payments for

    (a) transporting Equipment to and from the Working Areas
    (b) erecting and dismantling Equipment
    (c) upgrading or modification needed for a compensation event.

HORTER SCHEDULE OF COST COMPONENTS

This schedule is part of the *conditions of contract* only when Option A, B, C, D or E is used. When Option C, D or E is used, this schedule is used by agreement for assessing compensation events. When Option C, D or E is used, in this schedule the *Contractor* means the *Contractor* and not his Subcontractors. An amount is included only in one cost component and only if it is incurred in order to Provide the Works.

For main Options C, D and E this Schedule is only used by agreement between the *Project Manager* and the *Contractor* to assess compensation events. The default for these Options is the Schedule of Cost Components. Again, the extra note emphasises that the basis of administration of cost components is that they are incurred in order to Provide the Works.

People 1   The following components of the cost of

- people who are directly employed by the *Contractor* and whose normal place of working is within the Working Areas,
- people who are directly employed by the *Contractor* and whose normal place of working is not within the Working Areas but who are working in the Working Areas and
- people who are not directly employed by the *Contractor* but are paid for by him according to the time worked while they are within the Working Areas.

11   Amounts paid by the *Contractor* including those for meeting the requirements of the law and for pension provision.

ECC2 items 11 to 13 are consolidated into a brief but all inclusive statement that it is the 'amounts paid by the *Contractor*' which are used to assess the people costs.

Equipment 2   The following components of the cost of Equipment which is used within the Working Areas (including the cost of accommodation but excluding Equipment cost covered by the percentage for people overheads).

21   Amounts for Equipment which is in the published list stated in the Contract Data. These amounts are calculated by applying the percentage adjustment for listed Equipment stated in the Contract Data to the rates in the published list and by multiplying the resulting rate by the time for which the Equipment is required.

22   Amounts for Equipment listed in the Contract Data which is not in the published list stated in the Contract Data. These amounts are the rates stated in the Contract Data multiplied by the time for which the Equipment is required.

23   The time required is expressed in hours, days, weeks or months consistently with the list of items of Equipment in the Contract Data or with the published list stated in the Contract Data.

24   Unless the item is in the published list and the rate includes the cost component, payments for

- transporting Equipment to and from the Working Areas other than for repair and maintenance,
- erecting and dismantling Equipment and
- constructing, fabricating or modifying Equipment as a result of a compensation event.

25   Unless the item is in the published list and the rate includes the cost compone
       the purchase price of Equipment which is consumed.

26   Unless the item is in the published list and the rate includes the cost compone
       the cost of operatives is included in the cost of people.

Plant and Materials 3   The following components of the cost of Plant and Materials.

31   Payments for

       (a) purchasing Plant and Materials
       (b) delivery to and removal from the Working Areas
       (c) providing and removing packaging
       (d) samples and tests.

32   Cost is credited with payments received for disposal of Plant and Materials.

Charges 4   A charge calculated by applying the percentage for people overheads stated
       the Contract Data to the total of people items 11, 12 and 13 to cover the co
       of

       (a) overhead payments for people including payroll burdens
       (b) payments to utilities for the provision and use in the Working Areas
            water, gas, electricity and other services
       (c) payments to public authorities, utilities and other properly constitu
            authorities of charges which they are authorised to make in respect
            the *works*
       (d) payments for financing charges (excluding charges compensated for
            interest paid in accordance with this contract), buying or leasing la
            compensation for loss of crops or buildings, royalties, inspection cert
            cates, rent of premises in the Working Areas, charges for access to t
            Working Areas, facilities for visits to the Working Areas by Others a
            specialist services
       (e) payments for accommodation, equipment, supplies and services f
            offices, drawing office, laboratories, workshops, stores and compoun
            labour camps, cabins, catering, medical facilities and first aid, recreatic
            sanitation, security, copying, telephone, telex, fax, radio, CCTV, surve
            ing and setting out, computing, hand tools and hand held powered too

25   Unless the item is in the published list and the rate includes the cost component, the purchase price of Equipment which is consumed.

26   Unless included in the rate in the published list, the cost of operatives is included in the cost of people.

27   Amounts for Equipment which is neither in the published list stated in the Contract Data nor listed in the Contract Data, at competitively tendered or open market rates, multiplied by the time for which the Equipment is required.

> In a similar change to the Schedule of Cost Components, the cost of accommodation is now a recoverable cost under the heading Equipment. The previous ECC2 item 4(e) 'accommodation' provision has been deleted to reflect this change.
>
> ECC2 item 23 idle and standby times allowances for deduction have been deleted.

Plant and Materials 3   The following components of the cost of Plant and Materials.

31   Payments for

- purchasing Plant and Materials,
- delivery to and removal from the Working Areas,
- providing and removing packaging and
- samples and tests.

32   Cost is credited with payments received for disposal of Plant and Materials unless the cost is disallowed.

Charges 4   The following components of the cost of charges paid by the *Contractor.*

41   A charge calculated by applying the percentage for people overheads stated in the Contract Data to people item 11 to cover the costs of

- payments for the provision and use in the Working Areas of water, gas and electricity,
- payments for buying or leasing land, compensation for loss of crops or buildings, royalties, inspection certificates, charges for access to the Working Areas, facilities for visits to the Working Areas by Others and
- payments for equipment, supplies and services for offices, drawing office, laboratories, workshops, stores and compounds, labour camps, cabins, catering, medical facilities and first aid, recreation, sanitation, security, copying, telephone, telex, fax, radio, CCTV, surveying and setting out, computing, and hand tools not powered by compressed air.

42   Payments for cancellation charges arising from a compensation event.

43   Payments to public authorities and other properly constituted authorities of charges which they are authorised to make in respect of the *works.*

44   Consumables and equipment provided by the *Contractor* for the *Project Manager*'s and *Supervisor*'s office.

45   Specialist services.

> ECC2 item 4(a) 'overhead payments for people including payroll burdens' has been deleted from items to be included in the percentage for people overheads. This is covered in the revised ECC3 item 11. Again, payments for 'financing charges' has been removed along with 'accommodation'. The latter is recoverable as Equipment whereas the former is treated as being included in the Fee.
>
> Items 42 to 45 are now separately recoverable costs, not through the percentage for people overheads, items 42 and 44 reflecting similar provisions in the Schedule of Cost Components.

Manufacture and fabrication	5	The following components of the cost of manufacture or fabrication of Pla and Materials, which are

- wholly or partly designed specifically for the *works* and
- manufactured or fabricated outside the Working Areas.

	51	The total of the hours worked by employees multiplied by the hourly rat stated in the Contract Data for the categories of employees listed.
	52	An amount for overheads calculated by multiplying this total by the percenta for manufacturing and fabrication overheads stated in the Contract Data.
Design	6	The following components of the cost of design of the *works* and Equipme done outside the Working Areas.
	61	The total of the hours worked by employees multiplied by the hourly rat stated in the Contract Data for the categories of employees listed.
	62	An amount for overheads calculated by multiplying this total by the percenta for design overheads stated in the Contract Data.
	63	The cost of travel to and from the Working Areas for the categories of emplo ees listed in the Contract Data.
Insurance	7	The following are deducted from cost:

- costs against which this contract required the *Contractor* to insure and
- other costs paid to the *Contractor* by insurers.

Manufacture and fabrication	5	The following components of the cost of manufacture and fabrication of Plant and Materials, which are

- wholly or partly designed specifically for the *works* and
- manufactured or fabricated outside the Working Areas.

	51	Amounts paid by the *Contractor*.

A change from ECC2 is that the cost of manufacture and fabrication of Plant and Materials is now 'amounts paid by the *Contractor*' rather than the ECC2 approach of hours, hourly rates and adding a percentage for overheads. This measure, along with others, simplifies use of this schedule further.

Design	6	The following components of the cost of design of the *works* and Equipment done outside the Working Areas.
	61	The total of the hours worked by employees multiplied by the hourly rates stated in the Contract Data for the categories of employees listed.
	62	An amount for overheads calculated by multiplying this total by the percentage for design overheads stated in the Contract Data.
	63	The cost of travel to and from the Working Areas for the categories of design employees listed in the Contract Data.
Insurance	7	The following are deducted from cost

- costs against which this contract required the *Contractor* to insure and
- other costs paid to the *Contractor* by insurers.

This is the index to ECC3

Index by clause numbers (Option clauses indicated by their letters, main clause heads by bold numbers). Terms in *italics* are identified in Contract Data, and defined terms have capital initial letters.